Environmental Science and Engineering

Series editors

Ulrich Förstner, Hamburg, Germany
Wim Salomons, Haren, The Netherlands
Rod Allan, Burlington, ON, Canada
W.H. Rulkens, Wageningen, The Netherlands

For further volumes:
http://www.springer.com/series/7487

Environmental Science and Engineering

D. Padmalal · K. Maya

Sand Mining

Environmental Impacts and Selected Case Studies

Springer

D. Padmalal
K. Maya
Centre for Earth Science Studies
Thiruvananthapuram
Kerala
India

ISSN 1863-5520 ISSN 1863-5539 (electronic)
ISBN 978-94-024-0078-6 ISBN 978-94-017-9144-1 (eBook)
DOI 10.1007/978-94-017-9144-1
Springer Dordrecht Heidelberg New York London

Printed on acid-free paper

Springer is part of Springer Science+Business Media (www.springer.com)

Preface

Expanding human requirements and economic developments impose ever increasing pressure on the natural resource base. Advances in ecological and environmental studies reveal the imminent need for delimiting human aspirations and requirements within the barest minimum levels in order to achieve sustainability in developmental process.

For centuries, humans have been enjoying the natural benefits provided by rivers without understanding much on how the river ecosystem functions and maintains its vitality. Rivers are one of the important life sustaining systems of tropical and subtropical regions. Unlike large rivers, the small rivers of the world are the first to be hit by human interventions and/or economic developments. This is mainly because, the environmental subsystems in small rivers are closely knitted so that the areas available for dissipation of the adverse effects of the developmental processes are very limited. Among the various types of human interventions in river ecosystems, indiscriminate mining of sand and gravel is the most disastrous activity as it threatens the very existence of these systems. Depending on the geologic and geomorphologic setting, the degree of off-site and on-site impacts of sand mining would also vary. Continued and indiscriminate sand mining not only changes the physical characteristics of the river basin environments, but also disturbs its closely linked flora, fauna, and human life.

Sand in the river channel and floodplains constitute an important raw material for the construction industry. Sand production, movement and deposition are of great concern to engineers, geologists, and to geomorphologists, especially those concerned with river basin management, shore erosion, and harbor development. A better understanding of sand budget is very essential for solving problems concerning river and coastal environments. Besides its direct economic importance, sand also constitutes an important abiotic component in aquatic ecosystems like rivers. It provides suitable substrate for many benthic organisms. Sand is an unavoidable component for many organisms like fishes as it provides conducive environmental conditions for their breeding, spawning, feeding, hiding, etc. Further, inter-beds of sand within floodplain deposits act as aquifer systems storing large quantities of ground water for maintaining the base flow through the rivers. Sand acts as an efficient filter for various pollutants, and thus improves the quality of water in rivers and other related aquatic ecosystems. A cursory glance of the available studies reveal that it is nearly difficult to restore fully the structure and functions of the river ecosystems to their pre-disturbed state as the adversities of human interventions are alarming and mostly permanent. However, efforts are being made by the authorities concerned to regulate the activities within the resilience capability of the ecosystems. Lack of adequate knowledge on the different dimensions of the activity is often a major lacuna challenging the regulatory systems. This text aims to address a few aspects of environmental consequences of sand mining by taking the case of the small rivers draining the western flanks of the Western Ghats mountains as an example.

The entire work is addressed in 11 chapters. Chapter 1 deals with introduction on river sand and gravel mining, the functions of sand in natural river environment and its geologic origin. Chapter 2 summarizes the complex functions of river systems giving emphasis to the classic concepts in riverine studies. The different types of mining and mining methods used for extraction of sand from instream and floodplain areas are highlighted in Chap. 3. Chapter 4 is devoted to the environmental impacts of sand and gravel mining on different environmental components of river ecosystems. A review of the river sand mining scenario of the world with special emphasis to small rivers is given in Chap. 5. Chapter 6 deals with environmental case studies from Kerala State in South West India. The major drivers of sand and gravel extraction processes are also attempted in the light of the available literature. The environmental impacts arising from sand mining processes are assessed using two EIA methods—Matrix Method and Rapid Impact Assessment Matrix—in one of the river basins of Kerala and are presented in Chap. 7. In view of the various adversities of river sand mining, a detailed account of sustainable mining strategies and management measures to protect river systems are given in Chap. 8. Chapter 9 deals with sand auditing of Kerala rivers—a concept developed by the authors for assessing instream sand mining after a specific period of allowing sand mining. The properties of fine aggregates are discussed in Chap. 10. The last chapter (Chap. 11) deals with the various sources and conservation measures of sand. All the chapters are illustrated and supported using field evidences and available literature on the subject.

We believe, this text will be useful to students, researchers, decision makers and the general public, at large.

Thiruvananthapuram, India
April 2014

D. Padmalal
K. Maya

Acknowledgments

This book has evolved essentially from one and a half decade-long observations and studies of the authors in the small rivers draining the southwestern flanks of Western Ghat mountains in the western coast of India. Our contact and exchange of information with many experts in various disciplines related to the physical, chemical, and biological aspects of rivers helped immensely for the preparation of this book in the present form.

We are greatly indebted to the former Directors of the Centre for Earth Science Studies (CESS), Dr. K. M. Nair and Dr. M. Baba, and Prof. P. Seralathan (Rtd.), Cochin University of Science and Technology (CUSAT) for their constant encouragements and support throughout the course of this book assignment with Springer.

We thank Dr. N. P. Kurian, Director, Centre for Earth Science Studies (CESS) and Dr. C. N. Mohanan, Head, Environmental Sciences Division, CESS for facilities, support, and encouragements.

We greatly acknowledge Dr. S. Sreebha and Dr. P. R. Arun for their contributions in strengthening our understanding on river sand mining. We also thank our former Project Staffs Mr. Romio R. S., Mr. Sreejith S., Dr. S. Sajikumar, Dr. A. Krishnakumar, Dr. S. Sureshkumar, Dr. Rajesh Reghnath, Mrs. Mini S. R., Mrs. Sreeja S. R., Mrs. Ramya S. I., Mrs. Reshmi R., Mrs. Anooja S., Dr. R. S. Baiju, Dr. B. Baijulal, Dr. V. Santhosh, and Dr. Jissy Jyothi for their invaluable contributions while working at CESS in different sand mining projects.

With heartfelt gratitude, we remember Dr. K. K. Ramachandran (late), Former Advisor to Director, CESS for his constant encouragement during the initial phase of the sand mining studies in CESS. We greatly acknowledge Dr. S. P. Rajagopalan, Head, Computer Applications Division, Centre for Water Resource Development and Management (CWRDM) for his guidance and encouragement.

Officers of the Central Water Commission (CWC), Government of India and Irrigation department, Government of Kerala are acknowledged for providing data on water and sediment discharges as well as river cross profile data. Special thanks are also due to Presidents, Secretaries, and office staffs of Local Self Governments of Kerala and officers of the Revenue department, Government of Kerala, for all their help during field work and data collection. We thank Mr. N. K. Sukumaran Nair, General Secretary, Pampa Parirakshana Samithi (PPS), Pathanamthitta, for the help in various ways during the implementation of sand mining studies of Kerala rivers.

We thank Mr. Vishnu Mohan S., Ms. Rajimol T. R., Mr. Baburaj B., Mr. Ramdev P. R., Ms. Shiekha E. John, and Ms. Aswathy K. A. for their help in various ways during the preparation of this text.

We thank Prof. X. X. Lu, Department of Geography, National University of Singapore, Prof. Colin Kipatrick, Institute of Development Policy and Management, University of Manchester, United Kingdom, Dr. N. K. Tamrakar, Managing Director, Central Department of Geology, Tribhuvan University, Nepal, KSCSTE, Government of Kerala and Springer for granting permission to use some of the copy right materials. Prof. Ranjana U. K., Department of Geography, University of Colombo, Sri Lanka is acknowledged for his timely help in providing some inputs from the sand mining sector of Sri Lanka.

We are indebted to Springer for making this work a reality. Special thanks are due to Dr. Elodie Tronche, Ms. Naomi Portnoy, and Mr. Jan Marten de Vries of Springer Publishing Company for their encouragement and support during the preparation of this text.

Finally, we thank all those who are directly or indirectly involved in the work.

Thiruvananthapuram, India D. Padmalal
April 2014 K. Maya

Contents

Figures

Plates

Tables

Chapter 1
Introduction

Abstract Rivers, the most important life sustaining systems in tropics and subtropics, are under immense pressure due to various kinds of human interventions. Among these, indiscriminate sand and gravel mining from the active channels and floodplains are the most disastrous one as the activity threatens the very existence of the river ecosystem. The problem is acute in small rivers that have limited catchment area (<10,000 km^2) and river sand and gravel resources. Lack of adequate information on the adverse impacts of river bed resource extraction is a major setback challenging regulatory systems for the wise-use and management of the rivers and its finite resources. This chapter deals with the historical perspectives, environmental issues and ecological significance of the small rivers of the world that are more responsive to uncontrolled sand and gravel extractions.

Keywords River sand mining · Origin of sand · Uses of river sand · Small rivers

1.1 Introduction

Sand and gravel are the most important aggregate materials for building constructions. The ever increasing population and economic developments impose an exponential rise in their demand throughout the world (Kondolf 2000; Harrison et al. 2005; Lu et al. 2007). As sand and gravel can be extracted easily from river sources and such deposits do not require much processing other than size grading, most of the tropical and subtropical countries still depend on river sources such as instream, floodplain, and terrace deposits, to meet their aggregate requirements, especially sand—the fine aggregate—in construction works. But it is now well understood that continued and indiscriminate sand mining can cause irreparable and irreversible damages to the ecological and socioeconomic environments of the region, in the long run (Walker 1994; Kitetu and Rowan 1997; Kondolf 1998, 2000; Lu et al. 2007; Padmalal et al. 2008 and many others). The in-channel (from active

D. Padmalal and K. Maya, *Sand Mining*, Environmental Science and Engineering,
DOI: 10.1007/978-94-017-9144-1_1,

channel) and near-channel (floodplain and terrace) mining of sand inevitably alters the sediment budget of the river, in addition to considerably changing the channel hydraulics (Bull and Scott 1974; Erskine et al. 1985) and productivity of the system (UNEP 1990). Degradation of river environment due to sand mining could lead to serious environmental hazards including habitat transformations, biodiversity alterations and damage to civil construction structures attached to the river environment (FAO 1998). In most cases, lack of adequate information regarding adverse effects of sand and gravel mining is a major lacuna challenging the existing regulatory measures and also the management of the activity within the natural resilience capability of the river ecosystems. Further, scientific literature on the environmental impacts of sand mining in peer-reviewed journals and volumes are also scarce in properly addressing the gravity of the issue at different levels. And whatever available is often inadequate to meet the scientific requirements fully. Therefore, in this work, an attempt has been made to gather the available literature on sand and gravel mining in order to address the different dimensions of the problem especially with regards to environmental degradation and impact assessment with an aim to chalk out certain environment-friendly management strategies to salvage the rivers from indiscriminate sand and gravel mining.

1.2 Mining: The Historical Perspective

Mining and agriculture are the earliest activities of man and are the fundamental requirements in the development of civilized societies. The origin of mining dates back to the era of human evolution. Our ancestors have excavated flint and other minerals, rock fragments, etc., for making weapons for hunting, sizing the collected materials, and rescuing themselves from wild animals. Since then, the dependence of man on minerals and building materials have increased as the society evolved through the ages. Now, a stage has reached where the society cannot exist without the products of mining and quarrying and its demand will continue to grow in future also.

In the past, the mining sector was least concerned about the environmental repercussions of resource extraction. Today, the scenario has changed significantly as the stakeholders concerned have started giving more importance to the environmental impacts of resource extraction activities. This change in attitude, in turn, has lead to bringing in stringent regulations on mining and quarrying activities of all the natural resources. Many countries have brought out policies and legislations to lessen the impact of mining on various environmental components of the biological system. Planning for development, exploration, and conservation of mining are the three important issues that need to be addressed for the better management of the mining sector on one hand and conservation of minerals on the other.

1.3 River Sand Mining: Environmental Issues

For hundreds of years, sand and gravel are being used as aggregate materials for the construction of roads, buildings, and other civil works. As a result, the demand of sand is also rising exponentially in tune with the expansion of transportation and construction infrastructures. Sand and gravel are extracted from river channels for different purposes and for a variety of reasons—(a) for improving navigation, (b) for developing agricultural drainage systems, (c) for regulating flood, (d) for the production of aggregate materials, and so forth. Extraction of these materials for aggregates is the largest mining industry not only in terms of quantity but also in terms of revenue (Kondolf et al. 2002).

Despite the fact that sand is renewable in the geologic time periods, it is considered a nonrenewable resource as its regeneration is meager in the human calendar years. As the sand and gravel resources are extracted easily from the in-channel or near-channel sources, people depend on the river sources of sand greatly compared to the other aggregate sources. This has altered considerably the river systems (Plate 1.1) and the channel hydraulics in addition to the reduction of productivity within the in-channel and near-channel areas. A stage has now reached to evolve sustainable strategies to ameliorate the ill effects of river sand and gravel mining within the natural environmental equilibrium.

1.4 Sand: Ecological Values and Uses

The entire processes operating within a river ecosystem works on energy. When water flows from higher elevations to lower levels, its energy will be utilized for geological and geomorphological works that are in operation over millions of years. The energy dissipation occurs in river environments in many ways, which is explained in detail by Kondolf et al. (2002). The important means of energy dissipation processes in a river environment are (a) turbulence at steps in the river profile, (b) functional resistance of cobbles and boulders, (c) functional resistance of vegetation along bank, (d) in meanders/bends, (e) in irregularities of the channel bed and banks, and (f) transport of sediments downstream. As sand is the most common sediment type in river channels, a substantial part of the energy of river water will be used for the transport of sand in different modes—through bed, partly in bed and partly in suspension and fully in suspension under different energy conditions.

Natural sand is a unique particle aggregate which is widespread irrespective of the latitudinal and longitudinal constraints. When the hydrosphere and atmosphere that are so close to the solid earth is set in motion with vigor, sand particles begin to be in action by mixing in the respective fluids (air or water) in motion. This enhances the density of the fluid-mix, losing the vigor of flow of the fluid substantially. In other words sand particles play an important role in regulating the

Plate 1.1 Rivers and river sand mining. **a** River sand deposits in the in-curve of a river; **b** Indiscriminate sand mining from in-channel areas

velocity of flow of geological agents like wind and water, thereby protecting the biological richness and diversity to a great extent. In addition to the environmental and ecosystem services, sand has many other uses to mankind. Sand is a most common and widespread abrasive. Siliceous sand is a raw source for the manufacture of sodium silicate, carborundum (silicon carbide), common and optical glasses, and silicate bricks. Also, sand is a fundamental ingredient in mortar and concrete. It is added to clays to reduce shrinkage and cracking in brick manufacture.

Sand is mixed with asphalt in road dressing (Pettijohn et al. 1973) and is also used for foundry application, filtration for locomotives as friction sand, etc. Some sand deposits are explored for gold (alluvial placers), gems, platinum, tin, tungsten, monazite, zircon, rutile, ilmenite, garnet, etc. In addition, sand in the sedimentary layers form important reservoirs for the storage of valuable fluids (fresh water brines, petroleum and natural gas). Sand aquifers form excellent conduit for ground water movement.

1.5 Origin of Sand

Natural sands are weathered or worn out particles of rocks (Plate 1.2). Many minerals will be lost or modified during the weathering processes in the source area. According to Goldich (1938), the order of stability of minerals is the reverse order of crystallization in magmatic chamber (Fig. 1.1). Quartz is the last mineral to crystallize in the magmatic chamber, but is the stablest mineral in the weathering series. The earlier formed minerals like olivines, pyroxenes, and plagioclase feldspars will be easily altered to secondary products/minerals during chemical weathering. A significant portion of the weathered products will be later removed during erosion, and transported to the site of deposition. A part of the minerals will be changed during diagenesis as well (Pettijohn et al. 1973). Due to weathering of feldspars in the host rocks will be altered to kaolinite or an intermediate product; pyroxenes and amphiboles are more likely to dissolve and be transported as dissolved ions. Minerals like quartz are practically insoluble or sparingly soluble. Therefore, they will be left out in rivers and other depositional environments along with durable minerals like zircon, sillimanite, etc. Hence, such minerals remain almost unchanged in their chemical composition during weathering, erosion, transportation or deposition. River sands seem to be more feldspathic than either dune or beach sands. The ever changing climatic conditions, sea level positions and tectonic processes in the Pleistocene and Holocene epochs have had a major role in the formation of the present day sand deposits in different areas of the world.

1.6 Scope of the Present Study

The rivers of the world are undergoing severe degradation due to various kinds of anthropogenic activities like excessive utilization of river bed resources, discharge of domestic as well as industrial contaminants, input of chemical fertilizers from agricultural lands, construction of dams, etc. Among these, over exploitation of riverbed resources especially for construction grade sand and gravel is the most dangerous one as the activity adversely affects the existence of river ecosystems. In most cases, mining of river sand exceeds several folds the natural replenishments

Plate 1.2 The production of sand and gravel through hydraulic action (**a**) and (**b**) and abrasion (**c**) and (**d**)

Fig. 1.1 Mineral stability series in weathering (after Goldich 1938)

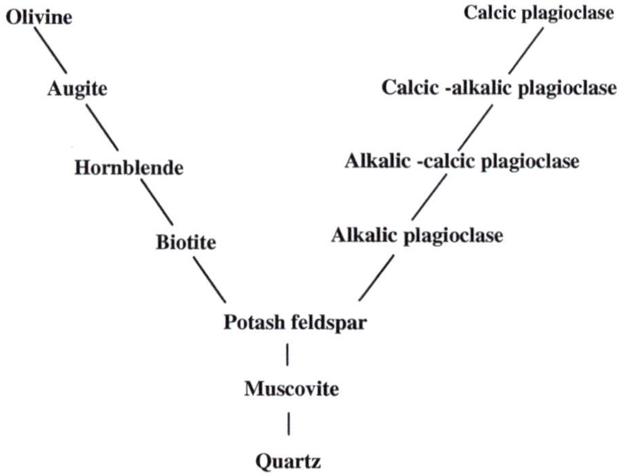

(Kondolf 1998). Unabated sand mining adversely affects the stability of riverbeds and banks, leading to degradation of the entire fluvial system. For conservation and sustainable utilization of resources, proper assessment has to be made on the availability of resources in the river channels and related alluvial sources. Also our understanding on the impacts of sand and gravel mining on various environmental components of the river systems have to be strengthened for taking timely decisions on matters related to river management. We hope the present work will be useful for chalking out new guidelines or regulations for protecting the river ecosystems of the world from various types of human interventions like river sand and gravel mining.

References

Bull WB, Scott (1974) Impact of mining gravel from urban stream beds in the Southwestern United States. J Geol 2:171–174

Erskine WD, Geary PM, Outhet DN (1985) Potential impacts of sand and gravel extraction on the Hunter river, New Southwales. Aust Geogr Stud 23:71–86

FAO (1998) Rehabilitation of rivers for fish. Food and agriculture. United Nations Organization. Fishing News Books, Oxford, p 260

Goldich SS (1938) A study of rock weathering. J Geol 46:17–58

Harrison DJ, Fidgett S, Scott PW, Macfarlane M, Mitchell P, Eyre JM, Weeks JM (2005) Sustainable river mining of aggregates in developing countries. In: Marker BR, Petterson MG, McEvoy F, Stephenson MH (eds) Sustainable minerals operations in the developing world. Geol Soc London Spec Publ 250:35–46

Kitetu J, Rowan J (1997) Integrated environmental assessment applied to river sand harvesting in Kenya. In: Patric CK, Lee N (eds) Sustainable development in a developing world—integrated socio-economic appraisal and environmental assessment. Edward Elgar, Cheltenham, p 189–199

Kondolf GM (1998) Large scale extraction of alluvial deposits from rivers in California: geomorphic effects and regulatory strategies. In: Klingeman PC, Beschta RL, Komar PD, Bradley JB (eds) Gravel bed rivers in the environment. Water Resources Publications, Colorado, pp 455–470

Kondolf GM (2000) Assessing salmonid spawning gravels. Trans Am Fish Soc 129:262–281

Kondolf GM, Smeltzer M, Kimball L (2002) Freshwater gravel mining and dredging issues. Report prepared for Washington department of fish and wildlife. Washington Department Of Ecology and Washington Department of Transportation, p 122

Lu XX, Zhang SR, Xie SP, Ma PK (2007) Rapid channel incision of the lower Pearl River (China) since the 1990s as a consequence of sediment depletion. Hydrol Earth Syst Sci 11:1897–1906

Padmalal D, Maya K, Sreebha S, Sreeja R (2008) Environmental effects of river sand mining: a case from the river catchments of Vembanad lake, Southwest coast of India. Environ Geol 24:879–889

Pettijohn FJ, Potter PE, Siever R (1973) Sand and sandstone. Springer, New York, p 618

UNEP (1990) Environmental guidelines for sand and gravel extraction projects. Environmental guidelines, No. 20, United Nations Environment Programme, Nairobi, p 37

Walker HJ (1994) Environmental impacts of river dredging in Arctic Alaska (1981–1989). Arctic 47:176–183

Chapter 2
Rivers-Structure and Functions

Abstract Rivers are the corridors connecting the terrestrial environment to the ocean realm. They play an important role in the sustenance of life systems of nature. As a geological agent, rivers carve out distinct suite of geomorphic features on the surface of the Earth. But human interventions consequent to economic developments in the past few decades have imposed tremendous pressure on rivers. As a result, most of the rivers in the world, especially the small rivers, have been altered to levels, often beyond their natural resilience capability. The present chapter gives a brief presentation of the river environment with special reference to its ecological and geological functions. River sediment characteristics, channel processes, classifications of rivers, and some of the classic concepts in riverine studies are also given in the chapter.

Keywords River ecosystem · Geological work of rivers · River sediments · Concepts in riverine studies

2.1 Introduction

Apart from being a crucial ecosystem linking the land and ocean systems, rivers serve as a prominent geological agent in tropical and subtropical regions. Running water is the most pervasive agent of erosion in nature. Under the influence of the force of gravity, rivers sculpture Earth's surface into distinct landforms (Leopold et al. 1964). The most striking example is the Grand Canyon that has been carved down by running water through a process that took millions of years. Like erosion of land, rivers play a major role in the transfer of materials from terrestrial environment to ocean realm (Lal 1977).

Milliman and Meade (1983) computed that $12–13 \times 10^9$ tons of suspended sediments are supplied to the world oceans annually by rivers and further $1–2 \times 10^9$ tons are supplied as bed load and flood water discharges. However, all sediments borne off by rivers seldom reach the sea as a considerable portion is

D. Padmalal and K. Maya, *Sand Mining*, Environmental Science and Engineering, DOI: 10.1007/978-94-017-9144-1_2,

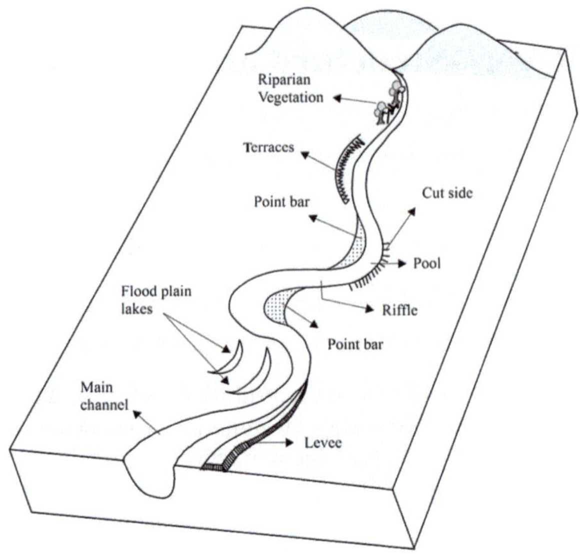

Fig. 2.1 Gross features of a river along the master channel

detained in the onland part of the fluvial system. Flowing water is the main agent responsible for the creation of physical habitat in a river environment (FAO 1998). Rivers create, destroy, and re-create distinct landforms and habitats through erosion, deposition, or a combination of these two. A low order stream in the uplands is usually erosional. This reach will essentially be composed of boulders, cobbles, and other coarser sediments, whereas the higher order rivers in the lowland contain sediments of silt-sand grades that are deposited in the form of bars, point bars, islands, natural levees, etc. A river system comprises the whole river corridor—the river channel, riparian zone, floodplain, and alluvial aquifer. Figure 2.1 depicts the gross features of a river along its master channel. Sand and gravel deposits constitute an integral part of this fluvial hydrosystem, which make it permeable to exchange of water between the corridors. In addition to this, sand and gravel dominant sediment substratum within river channels offer a conducive environment for many psammophilic organisms in the river environment. As the course of a river changes from upstream to downstream, the habitat and the communities in the environment would also vary. It is now well understood that the effects of alterations and perturbations in one stretch can be transferred into the downstream stretches located far from the source of the problem, inflicting the consequences at a distance (Jeffries and Mills 1990).

Despite its importance in supporting the life and greenery of the basin environment, rivers have been widely exploited by humans for natural resources—both living and nonliving, without understanding much the ecosystem functions (Naiman 1992; Naiman et al. 1995; Naiman and Bilby 1998). Studies reveal that man has changed the nature of many of the world rivers by controlling their floods, constructing large impoundments (Ittekkot and Lanne 1991), over exploiting the resources (Kitetu and Rowan 1997; Macfarlane and Mitchell 2003) and using rivers for disposal of wastes (Haslam 1990).

2.2 The River Ecosystem

River ecology deals mainly with the energy transformation, nutrient turnover, and storage and processing of organic matter. Rivers are basically heterotrophic as a substantial proportion of the biotic energy that drives stream communities is organic matter derived from allochthonous sources. Many aquatic plants, invertebrates, and fishes have adapted to fill a specific niche. Within most rivers, the pattern of flow variation, and its ramifications in terms of substrate stability and water quality, is the dominant factor controlling species distributions. Elwood et al. (1983) showed that lotic (pertaining to running water) ecosystems are longitudinally interdependent and that energy processing depends on the retention and cycling of nutrients by biological communities in upstream areas. From a biological point of view, flowing water has a number of advantages over still water. It constantly gets mixed up by turbulence providing nutrients, exchange of respiratory gasses, and removal of wastes. Lotic water is fundamental for the downstream and lateral movement of plants and animals. However, the character of flow changes from the headwaters to the river mouth, which in turn leads to a characteristic zonation in the riverine biological community. Biological community of a river ecosystem includes a variety of plants and animals. Producers in aquatic systems include diatoms, blue green algae, and water moss. Nymphs of dragon flies, may flies and stone flies, beetles, snails, fishes, etc. are the common consumers in river ecosystems. Some plants and animals can withstand the rapid flow of hill streams. Other species of plants and animals can live only in slow moving waters. Some species of fishes move to hill streams as they need crystal clear water to breed. The rising turbidity due to indiscriminate scooping of river bed materials for various purposes is an adverse condition for the existence of such fishes. In highly flooded rivers, recession of the annual flood delivers high levels of dissolved organic carbon and detritus (wood, leaves, seeds, etc.) to the main channel. This lateral connectivity is very important for sustaining the biological integrity of large rivers with well developed floodplains.

Riparian and instream vegetations are the integral components of the river ecosystems. The riparian vegetation plays an important role in sustaining the vitality of rivers. It is a source of organic matter, which forms an important source of energy in most of the river ecosystems. Further, the woody debris in aquatic ecosystems is an important habitat and spawning site for many aquatic animals. They not only play a pivotal role in stabilizing river banks from erosion, but also act as travel corridors for wild animals that connects with other ecosystems

Plate 2.1 Functions of riparian vegetation. **a** Dense riparian vegetation along the banks of river forms a protective sheath; **b** Riparian vegetation and its canopy not only protect river banks but also regulate water flow during high flow regime; **c** Bio-shielding of river banks by creepers; **d** Profuse development of roots of certain riparian plants (*Ochreinauclea missionis*) creates a blanket on river banks and thus rescue them from failure incidences/erosion; **e** and **f** The riparian vegetation provides habitat to many animals like Monitor lizard and birds

(Plate 2.1). The riparian environment hosts many ecologically significant and economically important plants (Plate 2.2). They offer shade and shelter to a variety of organisms including rare and endangered ones.

Ficus racemosa L.

Holigarna arnottiana Hook. f.

Ficus exasperata Vahl, Enum.

Hibiscus tiliaceus L.

Hydnocarpus pentandra (Bunch.-Ham.) Oken, Allg. Naturf.

Ochreinauclea missionis (Wall. ex G. don) Ridsd.

Plate 2.2 Some of the common riparian plants in the small rivers in the southwest coast of India

2.3 Geological Work of Rivers

The water flowing through a river could erode the land over which it flows, transport sediments that are formed by weathering and erosion, and finally deposit the transported materials, under favorable conditions into discrete landforms. According to Schumm (1977), an idealized river system can be divided broadly into three zones—(i) production zone or a zone of sediment erosion, (ii) zone of

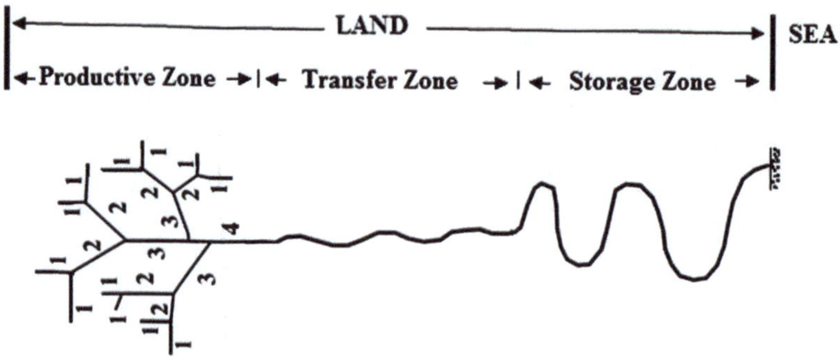

Fig. 2.2 The three primary zones of a river (after Schumm 1977)

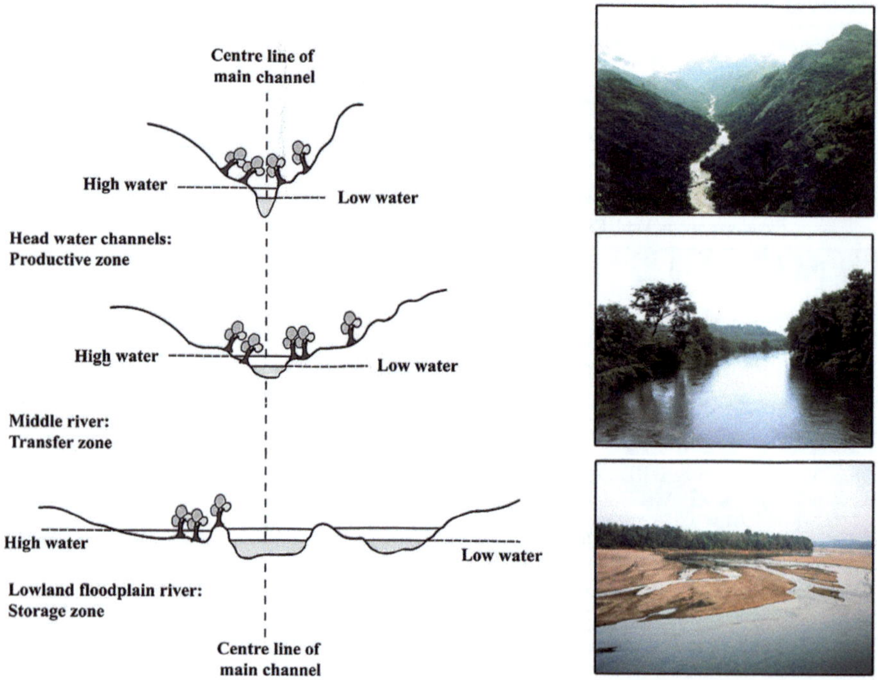

Fig. 2.3 Cross profiles of a river in the production, transfer and storage zones along with field evidences from the respective zones

sediment transfer, and (iii) a zone of deposition (Fig. 2.2). Figure 2.3 depicts cross profiles of a river in production, transfer, and storage zones along with field evidences from the respective zones.

The production zone will be steep, rapidly eroding head waters, whereas in transport zone the sediment is moved without net gain or loss. The transported materials will be deposited in the storage zone of the river under favorable conditions. The rate of erosion varies widely both spatially and temporarily. For example, the Appalachian mountains (North America) are eroded at about 0.01 mm/year (Leopold et al. 1964), the southern Alps of Newzealand at about 11 mm/year (Griffiths and McSaveney 1983), and southern central Range of Taiwan over 20 mm/year. The work of rivers is of considerable importance in land sculpturing. The entire processes involved in the geological work of rivers can be divided into three distinct processes, such as (a) erosion (hydraulic action, abrasion, attrition, and solution), (b) transportation (bedload, saltation, suspension, and dissolved forms), and (c) deposition (as fluvial landforms). Of these processes, the first process brings about degradation of the terrain, while deposition causes aggradation at suitable sites under favorable conditions. In the upper course of a river, processes are dominated by sediment production and incision of the channel into the landscape. The process of erosion becomes very conspicuous in excavating or downcutting the valley floor. The size of the sediment transported in any segment of the river is dependent on the geology of the basin as well as the distance of the segment from the source. The amount of sediment load carried depends on the size of the material, discharge, slope, and channel and catchment characteristics. When there is reduction in the discharge or in the slope of equilibrium, the same cannot transport the material supplied to it. Hence the stream deposits its excess materials in the form of floodplains, alluvial fans, deltas, etc., depending on the size and specific gravity of the material.

2.4 River Sediments and Channel Processes

River sediments comprise a spectrum of particle sizes such as boulder, cobble, pebble, granule, sand, silt, and clay (Lane 1947). Among these, the largest particles commonly occur in upland channels where the terrain gradient is the highest, while finer entities are enriched progressively downstream due to sediment sorting based on size and specific gravity (Blatt et al. 1972). Lower channel gradients in the lowlands favor deposition of small-sized particles and help floodplain development (Kondolf 1997). However, in the lower reaches, smaller tributaries can often contribute coarser particles to the mainstream. Table 2.1 shows the classification of sediment particles used widely by engineers and ecologists (Garde and Ranga Raju 1985; Jeffries and Miller 1990). Plate 2.3 shows the river bed characteristics of small rivers in the southwestern coast of India.

Collins and Dunne (1989) recognized three sediment delivery processes such as (1) mass wasting on hill slope, (2) hill slope erosion, and (3) erosion of channel bed and banks. Mass wasting includes landslides and soil creep, and occurs when gravity moves soil into the river channels. Hill slope erosion occurs when precipitation intensity exceeds the absorption capacity of soil and generates overland

Table 2.1 Grade scale for size terms applied for clastic sediments

Range in mm	Class name	Range in mm	Class name
4,096–2,048	Very large boulders	1/2–1/4	Medium sand
2,048–1,024	Large boulders	1/4–1/8	Fine sand
1,024–512	Medium boulders	1/8–1/16	Very fine sand
512–256	Small boulders	1/16–1/32	Coarse silt
256–128	Large cobbles	1/32–1/64	Medium silt
128–64	Small cobbles	1/64–1/128	Fine silt
64–32	Very coarse gravel	1/128–1/256	Very fine silt
32–16	Coarse gravel	1/256–1/512	Coarse clay
16–8	Medium gravel	1/512–1/1,024	Medium clay
8–4	Fine gravel	1/1,024–1/2,048	Fine clay
4–2	Very fine gravel	1/2,048–1/4,096	Very fine clay
2–1	Very coarse sand		
1–1/2	Coarse sand		

Source Lane 1947

flow. Stream channels and floodplains are built up and maintained by erosion and deposition of sediments during high stream flows (Whiting 1998). In relatively undisturbed river systems at its mature phase, gradual erosion of outside bends of meanders and deposition of eroded materials on inside bends cause an imperceptible shifting of channels within its floodplain. This form of stability is called dynamic equilibrium (Heede 1986). Although river flows and sediment loads vary annually or even seasonally, sediment balance and channel stability occur over the long-term. Instabilities introduced by human activities like deforestation, sand and gravel mining, and other activities, and natural processes like extreme precipitation, forest fires, and other events can cause channel bed and banks to become net sources of sediment.

2.5 River Classification

The classification of rivers are complicated by both longitudinal and lateral linkages, by changes that occur in the physical features over time, and by boundaries between apparent patches that are often indistinct. Davis (1899) divided rivers into three classes based on relative stages of channel developments—youthful, mature, and old age. River classification systems based on qualitative and descriptive delineations were developed by Melton (1957) and Matthes (1956). Broadly, two types of river classifications exist in literature—(1) physical and (2) biotic. Most classifications are based on characteristics of biota (Huet 1954; Hawkins et al. 1993) or valley types (Thornbury 1969) or fluvial features (Galay et al. 1973; Mollard 1973), but a few are based on other characteristics like levee formations (Culbertson et al. 1967) and floodplain types (Nanson and Croke 1992).

Plate 2.3 River bed materials. **a** and **b** Cobble bed river reach; **c** Rocky river bed; **d** Pebble bed—a close view; **e** Pebble bed showing patches of well sorted, fine to very fine sand deposited from bottom suspension due to obstruction by vegetation in the river bed; **f** Sand deposit in the storage zone

Another important classification proposed by Milliman and Syvitski (1992) is based on headwater elevation of rivers. They classified rivers as high mountain (headwater elevation >3,000 m amsl), mountainous (1,000–3,000 m amsl), upland (500–1,000 m amsl), lowland (100–500 m amsl) and coastal plain (<100 m amsl) rivers. The quality of river water was also considered, in some cases, for classifying rivers (Sioli 1950; Furch and Junk 1997). Based on this, the rivers are classified as black water rivers, white water rivers, and clear water rivers. A black water river is one with a deep, slow moving channel that flows through forested swamps and wetlands. The black color of river water is attributed to leaching of tannins from decayed leaves of adjoining vegetated lands. Black water rivers are seen in Amazonia, Orinoco basin, Southern and Northern United States. Negro river is the largest black water river in the world, and is one of the largest Amazonian tributaries. The white water rivers are muddy in color due to their high sediment content, while clear water rivers drain areas where there is little erosion. Based on the flow characteristics /water availability, rivers are classified into three—(1) ephemeral, (2) intermediate, and (3) perennial. Ephemeral rivers are those that do not flow continuously throughout the year. Such rivers are common in semi-arid and desert regions. Intermediate rivers are those that carry water almost throughout the year and dry up in extreme droughts. Some rivers flow throughout the year even in extreme droughts and are called perennial rivers.

2.6 Concepts in Riverine Studies

Ecological studies of river systems have made many advances during 1980s and 1990s, and several new paradigms such as the river continuum and nutrient spiraling concepts have emerged during this period. The following section deals with a brief account of the various classic concepts of riverine studies.

2.6.1 River Continuum Concept

The river continuum concept was developed by Vannote et al. (1980). The concept was developed from the idea that river systems, from headwaters to the mouth, present a continuous gradient of physical conditions. This gradient provides the physical template upon which the biotic communities and their associated processes developed. As a result, one would expect to observe recognizable patterns in the community structure and input, transport, utilization, and storage of organic matter. It proposes that energy inputs will change in a longitudinal direction, in relation to channel size, degree of shading, light penetration, and so on. And, the relative importance of functional feeding groups will change in tandem.

2.6.2 Nutrient Spiraling Concept

Webster (1975) and Webster and Patten (1979) proposed nutrient spiraling as a mechanism to account for the apparent ability of stream ecosystems to withstand and recover from disturbance. The term spiraling refers to the spatially dependent cycling of nutrients and the processing (i.e., oxidation and conditioning) of organic matter in lotic ecosystems. In effect, the nutrient spiraling concept provides both a conceptual and quantitative framework for describing the temporal and spatial dynamics of nutrients and organic matter in flowing waters. It also allows the structural and functional aspects of the biotic communities that enhance the retention and utilization of nutrients and organic matter to be interpreted in terms of ecosystem productivity and stability. Later Wallace et al. (1977) applied the idea in describing the role of filter feeders in streams. Newbold and his colleagues developed mathematical models at Oak Ridge National Laboratories to explain this concept (Newbold et al. 1981; 1982a, b; Elwood et al. 1983). These collective studies are termed as the nutrient spiraling concept.

2.6.3 Flood Pulse Concept

This concept was developed by Junk et al. (1989). The principal driving force behind the existence, productivity, and interaction of the diverse biota in river—floodplain systems is the flood pulse. A spectrum of geomorphological and hydrological conditions produces flood pulses, which ranges from unpredictable to predictable and from short to long duration. Short and generally unpredictable flood pulses occur in headwater streams and in streams heavily modified by human activities, whereas long duration and generally predictable floods occur in larger rivers. The net result is that the biota and the associated system level processes reflect the characteristics of the flood regime. Its basis is that the pulsing of river discharge—the flood pulse—is the major force controlling biota in river floodplains, and that lateral exchange between the river channel and its floodplains is more important in determining nutrient and carbon supply in lower reaches than longitudinal connections (Fig. 2.2).

2.7 Human Interventions

The dynamics of river systems are affected greatly by human interventions either within the catchment or directly within the river corridor (Petts and Calow 1996). River regulation by dams, diversions, channelization, and other physical controls, resource extraction including sand and gravel, etc., has significantly altered majority of the world rivers. The physical, chemical, and biological characteristics of rivers suffer from these effects. In addition to directly altering river flow, anthropogenic activities transform the landscape through which the river flows. Erosion delivers more sediment to river channel, with detrimental effects on the instream habitat. Removal of riparian vegetation leads to rise in temperature of overlying waters, shift from heterotrophy to autotrophy, reduced bank stability, and loss of the natural capacity to prevent sediments and nutrients from reaching

Plate 2.4 Environmental degradation caused by human interventions: **a** Disposal of solid waste into river channel is a common practice in many rivers; **b** Encroachment on a river channel; **c** and **d** Removal of riparian vegetation for sand mining and agricultural activities

river channels. Anthropogenic activities also enhance the quantity of chemical wastes entering in rivers, both from agricultural and urban sources.

Human impacts on river ecosystems may vary in character, and the way in which they are mitigated or resolved depends on whether they are 'planned' or 'unplanned' (Plate 2.4). 'Planned impacts' are usually direct and quickly felt effects, such as land use changes or in-channel changes by construction and /or mining activities. Clearing of forest cover on slopes close to channel margins may increase the quantity and speed of overland flow. Construction of bridges and check dams modifies local flow behavior in channels, causing scour of adjacent bed and bank sediments. Construction of large impoundments as well as large-scale riverbed mining is the most radical form of human impacts (Elliott and Parker 1997; Hadley and Emmett 1998 and Kondolf 1997). Although this form of intervention can exert long-term effects, most of them are, in principle, reversible, or at least amenable to mitigation. Channel works can be removed or redesigned; land use change can be reversed. 'Unplanned effects' are usually delayed in their onset, and are often more difficult to identify, and may be cumulative. Protection and restoration of rivers from anthropogenic disturbances is the need of the hour. Rivers have considerable ability to recover from 'pulse events', such as chemical inputs from an accidental spill or a point source. Unfortunately, threats like sand and gravel mining, channelization and other forms of regulations /modifications that affect morphological character of rivers are essentially continuous, or 'press events'. These require much more effort to address, and one hopes that the principles of ecology and geomorphology will be considered /viewed seriously while designing conservation and management strategies of these life-sustaining systems.

References

Blatt H, Middleton GV, Murray RC (1972) Origin of sedimentary rocks. Prentice Hall, New Jersey, p 634

Collins B, Dunne T (1989) Gravel transport, gravel harvesting, and channel-bed degradation in rivers draining the Southern Olympic Mountains, Washington, USA. Environ Geol Water Sci 13:213–224

Culbertson DM, Young LE, Brice JC (1967) Scour and fill in alluvial channels. U.S. Geological Survey, Open File Report, p 58

Davis WM (1899) The geographical cycle. Geogr J 14:481–504

Elliot JG, Parker RS (1997) Altered stream flow and sediment entrainment in the Gunnison gorge. J Amer Water Resour Assoc 33:1041–1054

Elwood JW, Newbold JD, Neill ORV, Winkle WV (1983) Resource spiraling: an operational paradigm for analysing lotic ecosystems. In: Fontaine TD, Bartell SM (eds) Dynamics of lotic ecosystems. Ann Arbor Science, Michigan, pp 3–27

FAO (1998) Rehabilitation of rivers for fish, food and agriculture. United Nations Organization, p 260

Furch K, Junk WJ (1997) Floodplain: ecology of a pulsing system. Springer, Berlin, p 126

Galay VJ, Kellerhals R, Bray DI (1973) Diversity of river types in Canada. In: Proceedings of hydrology symposium, fluvial process and sedimentation, National Research Council of Canada, pp 217–250

Garde RJ, Ranga Raju KG (1985) Mechanics of sediment transportation and alluvial stream problems, 2nd edn. Halsted Press, India

Griffiths GA, Mcsaveney MJ (1983) Hydrology of a basin with extreme rainfalls—cropp river, New Zealand. NZ J Sci 26:293–306

Hadley RF, Emmett WW (1998) Channel changes downstream from a dam. J Amer Water Resour Assoc 34:629–637

Haslam SM (1990) River pollution: an ecological perspective. John Wiley and Sons, Chichester, p 253

Hawkins CP, Kershner JL, Bisson PA, Brgant MD, Decker LM, Gregory SV (1993) A hierarchial approach to classifying stream habitat features. Fisheries 18:3–12

Heede BH (1986) Designing for dynamic equilibrium in streams. Water Resour Bull 22:351–357

Huet M (1954) Biologie, profiles en long et en travers des eaux courantes. Bulletin Francais de la peche et de la Pisciculture 175:41–53

Ittekkot V, Lanne RWPM (1991) Fate of riverine particulate organic matter. In: Degens ET, Kempe S, Richey JE (eds) Biogeochemistry of major world rivers. John Wiley and Sons, New York, pp 233–243

Jeffries M, Mills D (1990) Freshwater ecology: principles and applications. John Wiley and Sons, Chichester, p 252

Junk W, Bayley PB, Sparks RE (1989) The flood-pulse concept in river floodplain systems. Can Spec Publ Fish Aquat Sci 106:110–127

Kitetu J, Rowan J (1997) Integrated environmental assessment applied to river sand harvesting in Kenya. In: Patric CK, Lee N (eds) Sustainable development in a developing world—

integrated socio-economic appraisal and environmental assessment. Edward Elgar, Cheltenham, pp 189–199
Kondolf GM (1997) Hungry water: effects of dams and gravel mining on river channels. Environ Manage 21:533–551
Lal D (1977) The oceanic microcosm of particles. Science 198:997–1009
Lane EW (1947) Report of the subcommittee on sediment terminology. Trans AGU 28:125
Leopold LB, Wolman MG, Miller JP (1964) Fluvial processes in geomorphology. WH Freeman, San Francisco, p 522
Macfarlane M, Mitchel P (2003) Scoping and assessment of the environmental and social impacts of river mining in Jamaica. Warwick Business School, University of Warwick, p 86
Matthes G (1956) River engineering. In: Abbott PO (ed) American civil engineering practice, John Wiley and Sons, New York, pp 15–56
Melton RD (1957) Geometric properties of nature drainage system and their representation in E4 phase. J Geol 44:341–352
Milliman JD, Meade RH (1983) World-wide delivery of river sediment to the oceans. J Geol 91:1–21
Milliman JD, Syvitski JPM (1992) Geomorphic/tectonic control of sediment discharge to the ocean: the importance of small mountainous river. J Geol 100:525–544
Mollard JD (1973) Air photo interpretation of fluvial features. In: Fluvial processes and sedimentation, Research Council of Canada, Ottawa, pp 341–380
Naiman RJ (1992) Watershed management. Springer, New York, p 560
Naiman RJ, Bilby RE (1998). River ecology and management in the Pacific coastal ecoregion. In: Naiman RJ, Bilby RE (eds) River ecology and management: lessons from the Pacific Coastal Ecoregion, Springer, New York, p 1–22
Naiman RJ, Magnuson JJ, Mc Knight DM, Stanford JA (1995) The fresh water imperative. Island Press, Washington DC, USA, p 165
Nanson GC, Croke JC (1992) A genetic classification of floodplains. In: Brakenridge GR, Hagedorn J (eds) Floodplain evolution, geomorphology, p 459–486
Newbold JD, Elwood JW, O'Neill RV, VanWinkle W (1981) Measuring nutrient spiraling in streams. Can J Fish Aquat Sci 38:860–863
Newbold JD, Mulholland PJ, Elwood JW, O'Neill RV (1982a) Organic carbon spiraling in stream ecosystems. Oikos 38:266–272
Newbold JD, O'Neill RV, Elwood JW, VanWinkle W (1982b) Nutrient spiraling in streams: implications for nutrient limitation and invertebrate activity. Am Nat 120:628–652
Petts GE, Calow P (1996) River restoration. Blackwell Science Ltd., Oxford, p 270
Schumm SA (1977) The fluvial system. John Wiley and Sons, New York, p 338
Sioli H (1950) Das Wasser im Amazonasgebiet. Forsch Fortschr 26:274–280
Thornbury WD (1969) Principles of geomorphology. John Wiley and Sons, New York, p 618
Vannote RL, Minshall GW, Cummins KW, Sedell JR, Cushing CE (1980) The river continuum concept. Can J Fish Aquat Sci 37:130–137
Wallace JB, Webster JR, Woodall WR (1977) The role of filter feeders in flowing waters. Archi fuer Hydrobiologie 79:506–532
Webster JR (1975) Analysis of potassium and calcium dynamics in stream ecosystems on three southern Appalachian watersheds of contrasting vegetation. Ph.D Thesis (Unpublished), University of Georgia, Athens, U.S.A
Webster JR, Patten BC (1979) Effects of watershed perturbation on potassium and calcium dynamics. Ecol Monogr 49:51–72
Whiting PJ (1998) Floodplain maintenance flows. Rivers 6:160–170

Chapter 3
River Sand Mining and Mining Methods

Abstract Large-scale sand extraction from river environment for building constructions is a global phenomenon. Indiscriminate sand mining imposes a series of physical, ecological, and socio-environmental impacts on the river basins. Furthermore, the extraction of this granular material is unavoidable as the sustainability of construction industry depends heavily on it. In rivers, construction grade sand occurs in different sources—active channels, floodplains, and river terraces. Different mining methods (mechanical or manual) are adopted for the extraction of sand from these sources under dry (above water table) and/or wet (below water table) conditions. This chapter describes briefly the different sources, methods and hydrogeological bearings of sand and gravel extraction processes in river ecosystems.

Keywords Instream mining · Floodplain mining · Wet pit mining · Dry pit mining · Bar skimming

3.1 Introduction

On an ecosystem perspective, sand is an important abiotic component which provides habitat for many aquatic animals (Kondolf et al. 2002). Further, river channel sediments are crucial for building fluvial landforms. In the active part of the channels, sand and other sediment particles play a crucial role in reducing the hungry water effect of river waters in high flow regime. The granular particles constituting sand get mixed up well in river water as and when the energy regime increases above the level of competency of the sediment particles. Since the sediment-water mix has higher density than pure water, the net velocity of flow drops significantly which in turn reduces the damages caused by high energy river water, and thus protecting the river ecosystems from severe damages. Sand is a nonrenewable resource in human life scale. The resource can be extracted from river basins for various purposes with much ease. The impacts of sand mining

D. Padmalal and K. Maya, *Sand Mining*, Environmental Science and Engineering,
DOI: 10.1007/978-94-017-9144-1_3,

on various components of the river environment are not always obvious and visible immediately and hence have long been underestimated by researchers and river managers (UNEP 1990). However, in recent years, there has been a spurt of renewed activity to monitor and evaluate the environmental impacts of sand mining because of increasing awareness on the destructive effects of the activity in aquatic and terrestrial environments.

The environmental costs of sand extracted from the active channels, floodplains and/or terraces seldom figure in the cost-benefit analysis or environmental impact assessment of the extractive industries. This makes the river sources more profitable than other alternatives (Kondolf 1994a). Commercial sand extraction from river sources is a global phenomenon. Indiscriminate sand mining produces a series of physical, ecological, and socio-environmental effects. This is one side of the problem, whereas on the other side, the need for construction grade sand and gravel is mounting exponentially year after year. It is a fact that, the demand for sand has increased manifold in recent years in consonance with the expansion of transportation and construction infrastructure. Depending on the geological setting, sand mining can cause detrimental effects on the living environments. The sustainability and fate of fresh water resources for its specific utility is directly linked with the river health. But, the in-channel or near-channel sand extraction could inevitably alter the ecosystem balance as the process disturbs channel hydraulics and productivity of the nearby environments. Further, the process has considerable adverse effects on river bank stability, engineering structures constructed for river protection and the ability to meet societal requirements (Weeks et al. 2003; Padmalal et al. 2008). In short, the environmental degradation resulted from indiscriminate sand mining makes it difficult to provide the basic needs of the riparian communities (Starnes 1983; Rivier and Seguier 1985; Sandeecki 1989; UNEP 1990; Kondolf 1997; Brown et al. 1998).

3.2 Types of Alluvial Sand Extraction

In river environment, sand and gravel deposits are extracted mainly from two major sources: (1) active channels and (2) floodplain areas and overbank areas (like terraces). Figure 3.1 shows the fluvial geomorphic features encountered in a channel profile across the river in its storage zones (Kondolf 1994a, b). A few field examples of instream, floodplain and terrace mining are depicted in Plate 3.1.

3.2.1 Instream Mining

Extraction of sand and gravel from the active channel of a river is called instream sand mining. Instream (in-channel) sand usually requires less processing than any

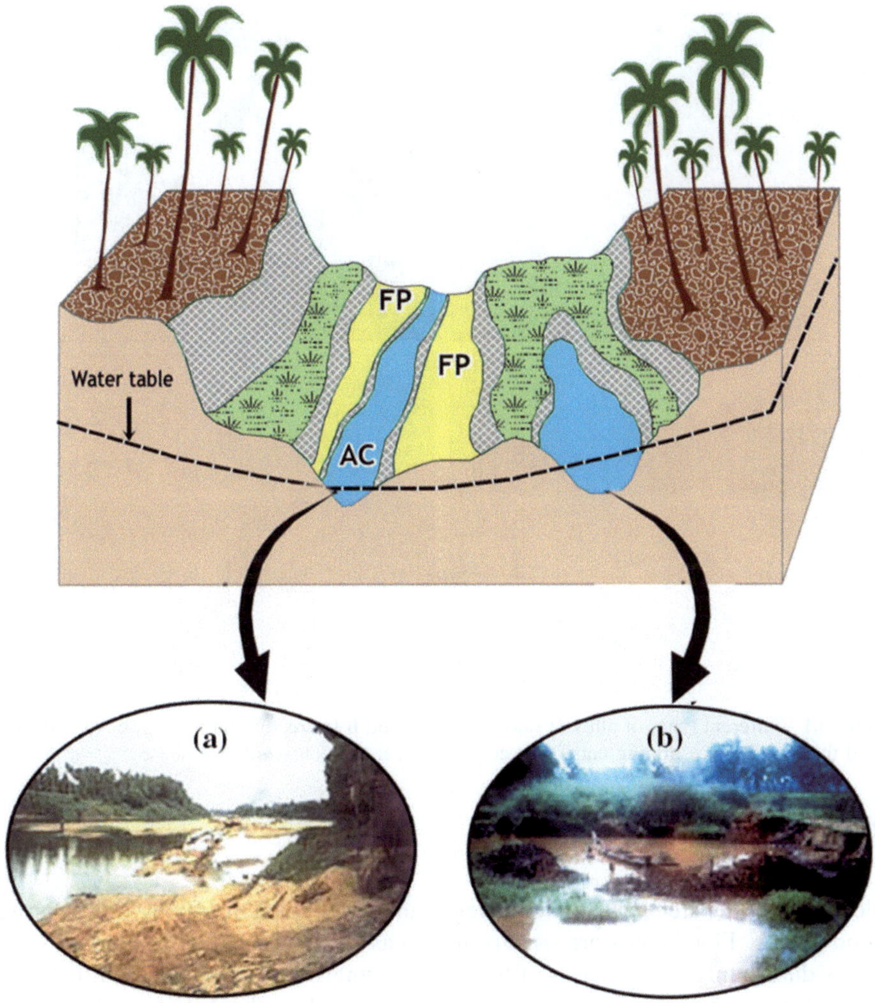

Fig. 3.1 A representative model showing instream (**a**) and floodplain (**b**) mining for sand and gravel from a river basin segment in the storage zone. *WT* water table; *FP* floodplain; *AC* active channel. The case of instream mining is from Periyar river and floodplain mining (wet pit mining) from Achankovil river (modified after Kondolf 1994b)

other sand sources. Instream sand mining takes place in many fluvial subsystems—bars, point bars or even active channels. Usually instream sand extraction takes place first followed by mining of sand from other alluvial sources.

Plate 3.1 Different types of alluvial sand mining. **a** and **b** Instream sand mining; **c** A floodplain sand mine; **d** Terrace mining under wet condition near the courtyard of a house

3.2.2 Floodplain Mining

Floodplain is the area just behind the levee and is occupied mainly by water during flood events. Floodplains are usually the areas evolved from deposition of sediments during migratory phases of the river channels. Stratigraphically, floodplain sediments are composed generally of channel sand at the bottom followed by floodplain silt and clay at the top. Mining of sand from the layer representing the channel sand in floodplain areas is referred to as floodplain mining (Fig. 3.1).

3.2.3 Terrace Mining

In this book, the term "terrace" is applied to the raised older river deposition areas seen behind the floodplains. In many rivers of the world, mining of sand from terraces is also very common.

3.3 Methods of Sand Extraction

A variety of methods are adopted for alluvial sand extraction. Kondolf et al. (2002) described several methods for the extraction of sand from the active channels and floodplain/terraces of riverine environments.

3.3.1 Active Channels

Instream sand and gravel mining (Plate 3.2) has been carried out by the following methods:

(a) *Bar scalping or skimming*
Bar scalping or skimming is the extraction of river bed materials, especially sand and gravel, from the top of the bars. The surface irregularities, if any, present in the sand bar will be smoothened by this process and the extraction of material will be limited to what could be taken above an imaginary line sloping upwards and away from water with respect to a specified level (Kondolf et al. 2002). Bar skimming is usually practiced every year based on the rate of deposition of sediments. To maintain the hydraulic control prevailed upstream by the riffle head, the preferred method of bar skimming is now generally to leave the top one-third of the bar undisturbed. Mining is confined only to two-third portion in the downstream end of the bar.

(b) *Dry pit channel mining*
Excavation of sand within the active channel of dry intermitted or ephemeral stream beds using mechanical (i.e., using bulldozers, scrapers, loaders, etc.) or manual means. Dry pits are often left with abrupt upstream margins (Plate 3.2). These abrupt margins act as head cuts during high flow seasons which will propagate upstream causing damages to the natural and man-made structures/features associated with river channels.

(c) *Wet pit channel mining*
Wet pit mining involves excavation of a pit in the active channel below the surface water of a perennial stream or below the alluvial groundwater table. In most cases, this sort of mining requires the use of a drag line or hydraulic excavator to extract sand and gravel from below the water surface.

(d) *Bar excavation*
A pit is excavated at the downstream end of the bar as a source of aggregate and as a site to trap sand and gravel. After completion of sand and gravel extraction, the pit may be connected to the channel at its downstream end to provide side channel habitat.

(e) *Instream sand and gravel traps*
Sand and gravel traps or bed load traps have been used to reduce sand movement in downstream channels for habitat enhancement. These traps can

Plate 3.2 Different methods of sand mining. Mechanical sand mining from ephemeral (**a**) and perennial (**b**) rivers; Manual mining from active channels (**c, d**); Manual mining from point bar deposit (**e**); Wet pit mining from floodplain (**f**)

be potential sources of commercial aggregate, provided the quantity so collected is economically viable. One advantage of the trap is the concentration of mining impacts at one site, where from the heavy equipments can remove sand and gravel without impacting much on the riparian vegetation or other natural channel features. Once the gravel traps are set up, sand and gravel extraction can be carried out on an annual basis.

(f) *Channel-wide instream mining*
Channel-wide instream mining is the extraction of sand and gravel from the entire active channel during the dry season. This kind of mining is practiced in rivers having variable flow regimes. The bed is evened out and uniformly lowered. This method is not promoted by many developed countries because of the concerns over the habitat impacts.

3.3.2 Floodplains and Terraces

Other important alluvial sources of sand are floodplains and river terraces. Two types of mining are being practiced in these source categories. They are (a) wet pit mining and (b) dry pit mining. In wet pit mining, the depth of the excavation pit crosses the groundwater table; whereas in dry pit mining, extraction of sand is limited to the upper dry bed.

3.4 Manual and Mechanical Mining

River sand mining is carried out both manually and mechanically. Manual mining is more environment-friendly and the quantity of mining is practically low. This method is practiced in many developing and underdeveloped countries having small rivers with limited river bed resources (Plate 3.2). Usually country boats and specially designed similar sand scoopers are used for extraction of sand in manual mining processes. In mechanical mining, high power jet pumps and heavy machineries are used for sand extraction from the active river channels and its floodplains.

References

Brown AV, Lyttle MM, Brown KB (1998) Impacts of gravel mining on gravel bed streams. Trans Am Fish Soc 127:979–992

Kondolf GM (1994a) Geomorphic and environmental effects of instream gravel mining. Landscape Urban Plann 28:225–243

Kondolf GM (1994b) Environmental planning in the regulation and management of instream gravel mining in California. Landscape Urban Plann 29:185–199

Kondolf GM (1997) Hungry water: effects of dams and gravel mining on river channels. Environ Manage 21:533–551

Kondolf GM, Smeltzer M, Kimball L (2002) Freshwater gravel mining and dredging issues. Report prepared for Washington department of fish and wildlife, Washington Department of Ecology and Washington Department of Transportation, p 122

Padmalal D, Maya K, Sreebha S, Sreeja R (2008) Environmental effects of river sand mining: a case from the river catchments of Vembanad lake, Southwest coast of India. Environ Geol 24:879–889

Rivier B, Seguier J (1985) Physical and biological effects of gravel extraction in river beds. In: Alabaster JS (ed) Habitat modification and freshwater fisheries. FAO, Rome, pp 131–146

Sandecki M (1989) Aggregate mining in river systems. Calif Geol Surv 42:88–94

Starnes LB (1983) Effects of surface mining on aquatic resources in North America. Fisheries 8:2–4

UNEP (1990) Environmental guidelines for sand and gravel extraction projects. Environmental guidelines, no. 20. United Nations Environment Programme, Nairobi, p 37

Weeks JM, Sims I, Lawson C, Harrison DJ (2003) River mining: assessment of the ecological effects of river mining in the Rio Minho and Yallahs rivers, Jamaica. British Geological Survey Commissioned Report, CR/03/162N, p 53

Chapter 4
Impacts of River Sand Mining

Abstract Mining of sand manifold higher than natural replenishments leads to severe damages to river systems. The mining process not only intercepts movement of sediments along the river channels, but disturbs the sediment balance established in the system over the geological time periods. Indiscriminate sand mining imposes several adverse impacts on the various environmental subcomponents of river ecosystems like bed forms, sediment milieu, water quality and quantity, flora and fauna, and socio-economic conditions of the people in the long run. The magnitude of these impacts depends on several factors such as type and scale of sand extraction, channel morphology, sediment transport processes and induced alterations in watershed characteristics. A better understanding of these impacts is very essential for formulating appropriate strategies for ameliorating and negating the impacts of river sand mining.

Keywords Channel incision · Bed degradation · Bed coarsening · Riparian vegetation · Pit capturing

4.1 Introduction

Mining of sand and gravel manifold higher than the natural replenishments may lead to irreversible and irreparable damages to land, water, and biotic components of the fluvial environment. The impacts of river sand mining will not be readily felt at measurable levels as it requires a decade or more to surface. Hence, rivers are said to possess "long memories," meaning the channel adjustments to instream extraction or comparable perturbations may persist long after the activity has practically ceased (Kondolf 1998). Sand mining disturbs the equilibrium that prevails in a river channel. The mining process intercepts movement of sediments and triggers an initial morphological response to regain the balance between supply and transport. Most studies on the geomorphological effects of river sand mining have involved extraction rates lasting over a decade or more, resulting in

D. Padmalal and K. Maya, *Sand Mining*, Environmental Science and Engineering,
DOI: 10.1007/978-94-017-9144-1_4,

large, measurable changes in channel geometry that are quite enough to be clearly detected. Although the direct ecological effects of resuspended sediments arising from sand mining sites are well documented, the long term, indirect effects of sand mining on the biological system, especially on the food web are often difficult to measure. Determination of site-specific impacts of sand mining needs careful and continuous assessments of mining-induced changes with respect to baseline conditions (Kondolf et al. 2002). The extent of these effects depends on many factors, such as the type and scale of sand extraction, resistance of the river channel to erosion, and changes in watershed characteristics such as hydrology, land use, and sediment transport processes. All this information is required in order to address the impacts of sand mining on the natural environments.

4.2 Changes in Bed Forms

A river channel is evolved into different bed form units depending on the changes in flow energy and sediment discharge through the system. Indiscriminate sand mining is one of the most destructive anthropogenic activities (Plate 4.1) hindering the natural stream bed evolution. A better understanding of the general distribution, sources, and fate of sediments is essential to discriminate the impacts of sand mining from the impacts caused by other types of natural and anthropogenic processes.

Rivers transport sediments and water from headwaters to its mouth. River channels are built up and maintained by erosion and deposition of sediments during river flows (Heede 1986; Whiting 1998). Current velocity exhibits significant variation in space and time. Most of the rivers experience a wide range of flows along and across its course. In relatively undisturbed river systems, a condition known as dynamic equilibrium exists where gradual erosion of outside bends of river meanders and deposition of eroded materials in the incurves take place (Fig. 4.1). Channel stability of a given river stretch is the outcome of a delicate balance existing between river flow, channel form, influx of sediment from the watershed, and loss of sediment to downstream reaches. This "conveyor belt" effect, where rivers transport eroded materials from headwaters to downstream stations provides the necessary quantity of sediment during channel forming flows such that channels remain in a dynamically stable condition (Kondolf 1997).

Instream mining results in channel instability through direct disruption of pre-existing channel geometry or through the effects of incision and related undercutting of banks (Collins and Dunne 1989; Erskine 2008). Mining aggravates widespread instability because the discontinuity in the sediment supply-transport balance tends to migrate upstream as the river bed is eroded to cover up the supply deficiency (Knighton 1984). Thus, sand mining from a relatively confined area triggers erosion of bed and banks, which in turn, increases sediment delivery to the site of original sediment removal. Bed degradation is caused by pit excavation and bar skimming, the two common methods of sand mining. Bed degradation occurs through two primary means: headcutting and "hungry" water effects. In the first

Plate 4.1 Selected scenes from Manimala river showing the extent of environmental degradation. **a** Ramps constructed for the passage of vehicles into the river bed; **b** A road constructed along the river channel for easy transportation of sand; **c** Bank erosion and uprooting of coconut trees due to unabated sand mining close to river bed; **d** Evidence of indiscriminate sand mining from prohibited areas close to the bridge; **e** River bed changes due to sand mining is a nuisance to local community that depends on the river for various purpose; **f** Remnants of ramps constructed for easy sand mining from the river channel has often become a nuisance to the river environment

case, excavation of a mining pit in the active channel lowers the stream bed, creating a nick point which attributes a local steepening on the channel system (Kondolf 1998). During high flows, the nick point becomes the foci of bed erosion which gradually moves upstream through head cutting (Bull and Scott 1974;

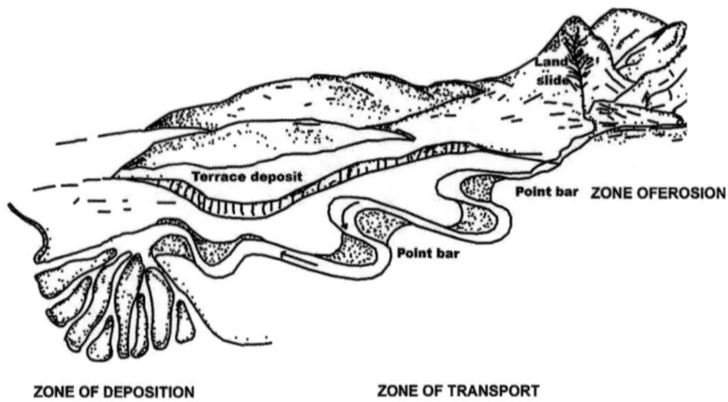

Fig. 4.1 Zones of erosion, transport, and deposition of a river system (modified after Kondolf 1997)

Hartfield 1993; Kondolf 1997). Head cutting generates substantial quantities of stream bed materials that will be transported further downstream to deposit in the excavated areas (Fig. 4.2). Head cuts often move long distances upstream and into the tributaries (Hartfield 1993; Kondolf 1997). Of the two common forms of bed degradation, head cutting is more recognizable in the field and exerts severe damages on the river environment. While pit excavation locally increases the channel depth by creating deep pools within the river channel, bar skimming widens the river channel. Both the conditions produce slower stream flow velocities. This in turn lowers the flow energy of river waters, causing the incoming sediments to deposit at the mining pit. As stream waters leave the area after deposition of sediments in the pit, the flow energy increases significantly in response to the "normal" channel flow. The amount of transported sediments leaving the site is now less compared to the sediment carrying capacity of the river water. Such energized river waters that emerge from the mining pit are often referred to as 'hungry water.' The hungry water accommodates its fury by eroding more materials from the stream reach below the mining site, causing bed degradation in the downstream areas. This condition continues till a balance between input and output of sediments re-establishes (Williams and Wolman 1984; Kondolf 1997). The interplay of these two processes of bed degradation and consequent channel incision will be visible even in the mining prohibited areas of bridges, water intake units, and other engineering structures that are constructed to protect river environments. Plate 4.2 gives a recent evidence of bridge collapse in Bharathapuzha river (SW India) near Shornur caused by channel incision.

The shape of the channel cross-section in a given site is a function of factors like flow velocity, the quantity and quality of the sediment in movement through the section, and the nature of the materials in the in-channel and off-channel areas. Sand mining not only reduces the stability of river banks but also adversely affects the riparian flora and fauna. The material comprising the river bank is acted upon mainly by two forces tending to produce downslope movements: (1) gravity,

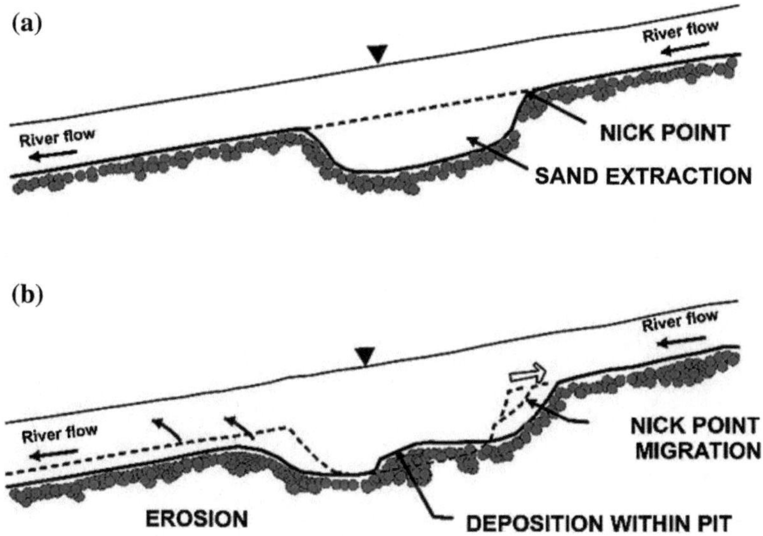

Fig. 4.2 Incision produced by instream sand mining. **a** Sand extraction creates a nick point; **b** The nick point migrates upstream and the hungry water emerging from the deposition of sediments in the excavation pit erodes the bed downstream (modified after Kondolf 1997)

which tends to make the material roll or slide downslope, (2) movement of water through the channel, which tends to drag or push the material. The magnitude of the force depends on the flow velocity, which in turn will be controlled generally by roughness of the streambed resulted in from pit excavation of sand and gravel. Bank erosion and bank retreat are frequently observed in river stretches undergoing indiscriminate sand mining. The river banks derive their strength and resistance to erosion largely from vegetation (Yang 2003) and to a lesser extent from composition, height, and slope. But channel incision due to indiscriminate sand mining and pit formation is a threat to even the otherwise stabilized river banks.

4.3 Changes in Sediment Characteristics

Indiscriminate and continued sand mining from alluvial reaches impose notable changes in the grain size characteristics of river sediments (Scott 1973; Sandecki 1989; Stevens et al. 1990). As bed materials form an important abiotic component of a river ecosystem, these changes often lead to changes in biodiversity of the river systems (Starnes 1983; Thomas 1985; Kondolf and Matthews 1993). Studies carried out in the Manimala and Muvattupuzha rivers in southwestern India reveals that indiscriminate river sand mining has imposed marked changes in the grain

Plate 4.2 Collapse of bridge in Bharathapuzha river (SW India) near Shornur due to channel incision

size spectral image of the in-channel sediments (Padmalal et al. 2010; Anooja et al. 2011). Analysis of sediments collected from selected point bars of the Manimala river shows bed coarsening consequent to the removal of finer particles from the upstream end of the bars. The continued operation of sand mining in the downstream areas ends up in bed coarsening of the upstream part of the point bars. Hence, point bars once composed entirely of sand-sized particles turn to gravelly sand and sandy gravel due to selective entrainment and removal of medium to very fine sands from the sediment population. The best example is the case of Kallooppara point bar which was once used for holding religious congregations. But now the upstream end of the point bar has turned gravelly sand (Fig. 4.3) because of the preferential enrichment of lag concentrates of coarser particles like pebbles and cobbles. The sand particles that are selectively removed from the upstream end of point bar later get deposited in the downstream end or in the deep pools resulted from pit excavation of sand. The unabated sand mining prevalent in rivers aggravate the natural rhythm of sediment sorting leading to marked changes in the textural fabric of the naturally evolved bars and point bars.

Table 4.1 shows a comparative evaluation of the mean size and standard deviation of bed materials computed for the Muvattupuzha river. Except one distributary channel, bed coarsening was noticed in all the other reaches of the Muvattupuzha river. The mean size in the tributaries varies from average -0.10Φ to -0.62Φ in 2010 against the previous report of 0.10Φ to -0.09Φ by Padmalal (1992). A similar trend is observed in the mean size of sediments in the main

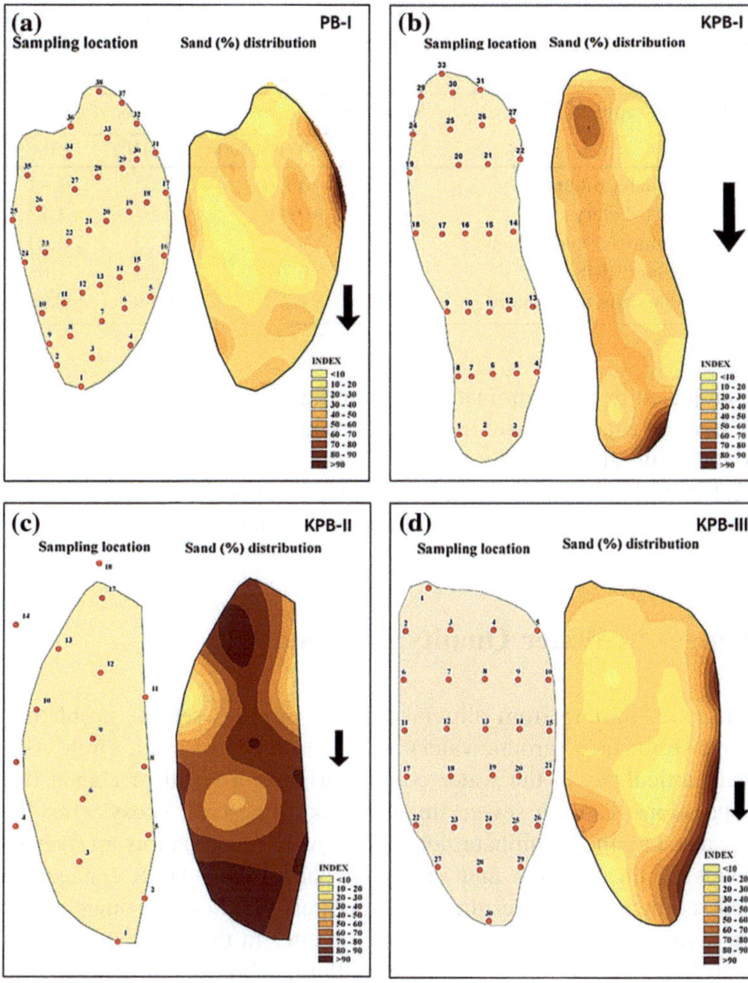

Fig. 4.3 Sand distribution map of some selected point bars in Manimala river. **a** Prayattu *kadavu* bar (PB-1); **b** Kallooppra point bar-I (KPB-I); **c** Kallooppara point bar-II (KPB-II) and **d** Kallooppara point bar-III (KPB-III); *Arrow* indicates the direction of river flow

channel (present mean size—0.18Φ; previous mean size—0.52Φ) and distributary channel (present mean size—0.67Φ; previous mean size—1.21Φ) as well. This clearly indicates the fact that sand mining in the river reach over the years is a major causative factor for the observed changes in the granulometric attributes of the sediment substratum. Sand mining activates the selective removal of medium to very fine sands from upstream and their deposition in the subsequent downstream stations. On the contrary, continued removal of construction grade sand from the downstream reaches and subsequent replacement by finer sands from upstream stations could attribute fining of grain size populations in the distributary

Table 4.1 Comparative evaluations of mean and standard deviation estimated for the tributary-main channel-distributary sediments of Muvattupuzha river with that of the earlier survey of Padmalal (1992)

Sl. No.	Channel segment	Mean (ϕ)		Standard deviation (ϕ)	
		(2010)	(1989–1990)	(2010)	(1989–1990)
1	Thodupuzha tributary	−0.10	0.1 0	1.65	1.41
2	Kaliyar tributary	−0.62	−0.09	1.64	1.73
3	Muvattupuzha main channel	0.18	0.52	1.22	1.44
4	Ittupuzha distributary	1.04	0.76	1.05	1.0
5	Murinjapuzha distributary	0.67	1.21	0.99	0.98

channel subjected to indiscriminate sand mining. Riverbed lowering will have an adverse effect on psammophilic fishes that require sandy substratum for spawning (Sheeba and Arun 2003). Further, lack of sufficient quantity of sand in the sediment substratum could enhance the hungry water effect (the accelerated destructive effects of sediment-deficit water) in rivers during high flow periods.

4.4 Changes in Water Quality and Quantity

Sand and gravel mining from alluvial reaches imposes serious problems in the surface and subsurface (groundwater) water resources as well. High content of suspended particulates in the water column arise as a result of clandestine sand mining operations, causing severe impairments to the river ecosystems. The disturbance caused by indiscriminate sand and gravel mining churns up river water to form clouds of fine organic and inorganic particulates. High concentration of suspended sediments block respiratory structures of fishes and other aquatic animals like bivalves. High load of particulate matter in the overlying water impair respiration and photosynthesis of instream flora, which in turn lead to reduced growth rate and finally its total destruction (Thrivikramaji 1986; UNEP 1990). Deposition of silt on river bed can smother diatoms, benthic algae, macro invertebrates, and fish eggs (Kanehl and Lyons 1992; Nelson 1993; Meador and Layher 1998; Lake and Hinch 1999).

Another problem that may arise in due course of sand mining operations is the imposed changes in river bed characteristics. The silt and clay particles arising from sand extraction will be carried downstream depending on the flow energy of the river, which then blankets the sand bars in the prohibited areas of the river channel. This leads to stabilization of sand bars by thick vegetative growth. Pure sand deposits (i.e., sand bars) cannot sustain vegetation. Plate 4.3 shows mud blanketing and stabilization of sand bars along the Periyar river in southwestern India which has resulted from clandestine sand mining over the past few decades in upstream reaches. Furthermore, particulate sediments are one of the efficient carrier phases of nutrients, the variations of which in river waters can alter the net

Plate 4.3 Imposed river bed changes noticed in Periyar river during 1998–2008 (a midland location) due to heavy siltation resulted from selective subaqueous extraction of construction grade sand using porous scoopers and subsequent stabilization of the instream sandbar by thick vegetation

nutrient flux to the nearshore marine environments. It is a fact that large pits in the lowland part of the river channels developed as a result of rampant sand mining often trap fine sediments loaded with major and minor nutrients. Pit excavation and trapping of nutrient loaded fine particulates, including the TSS within it, lead to reduction of nutrient supply from terrestrial environments to ocean realm (Babu and Sreebha 2004).

The Planning Department of the Lake County, California in 1992 reported that there is a potential reduction in alluvial aquifer storage in areas close to indiscriminate sand mining reaches (Lake County 1992). Kondolf et al. (2002) while reviewing the effects of sand and gravel mining on various environmental components opined that channel incision consequent to indiscriminate sand mining can accelerate the pace of groundwater lowering in areas adjacent to river channels. Indiscriminate extraction of huge volumes of sand from rivers flowing over deep alluvium and/ or other erodible materials could accelerate channel incision and riverbed lowering. This in turn can lower the groundwater table leading to adverse hydrologic effects in the nearby areas (Evor and Holland 1989; Goodwin et.al. 1992). Groundwater lowering can also kill vegetation in floodplain wetlands and also along sloughs, where trees play important role in providing cover, shade, and supply of wood for salmonid habitat.

Groundwater lowering can obstruct recharge of a stratigraphically high aquifer. This could in turn affect low-flow regime of rivers. Such hydrological changes have been reported from many parts of the world—Cache Creek, California (Wahler 1981; Collins and Dunne 1990), Russian River (Goodwin et.al. 1992), etc. When channel incision and bed lowering are large, overbank flooding is virtually eliminated (Collins and Dunne 1990). The reduction in overbank flooding can reduce supply of organic-rich fine particulates to the floodplain, and could reduce replenishment of water to floodplain wetlands and sloughs and aquifers. Reduction of overbank flooding can also aggravate downstream flooding because of the loss of floodplain water storage. Large scale bed lowering can increase bank height and aggravate bank erosion and tributary stream bed erosion.

Drinking water shortage has caused great concern in many parts of south-western India, as groundwater forms the major drinking water source in the region. A case study carried out by Prasad and Nair (2004) in Achankovil river basin in Kerala has highlighted that indiscriminate sand mining from the river channel and consequent channel incision are the major causative factors of drinking water problems in areas adjoining the river channels in the midlands and certain parts of the lowlands (Fig. 4.4). According to Prasad and Nair (2004), out of the 53 wells surveyed in the alluvial reach of the river, 36 % are perennial and 64 % are seasonal that are drying up for varying durations. The wells were classified according to their year of construction, i.e., <5 years, 5–10 years, 10–20 years, and >20 years. The maximum percentage of wells that go dry for various periods are the oldest ones. It may be noted that the wells are dug during dry months and they are bottomed in a zone that yields the required quantity of water. This obviously shows that the water table has been lowered. When a well goes dry the consequential action is to deepen it and obviously many wells under all categories have been deepened to the extent possible. It may also be noted that the percentage of wells that do not dry up is under the category of <5 years. These wells were dug when the riverbed had reached closer to the present level. Deepening of many older wells to the level of the present summer riverbed level is not practical due to the presence of massive basement or fear of collapse on account of caving in the zone of weathering. All these supplement to the fact that channel incision has a

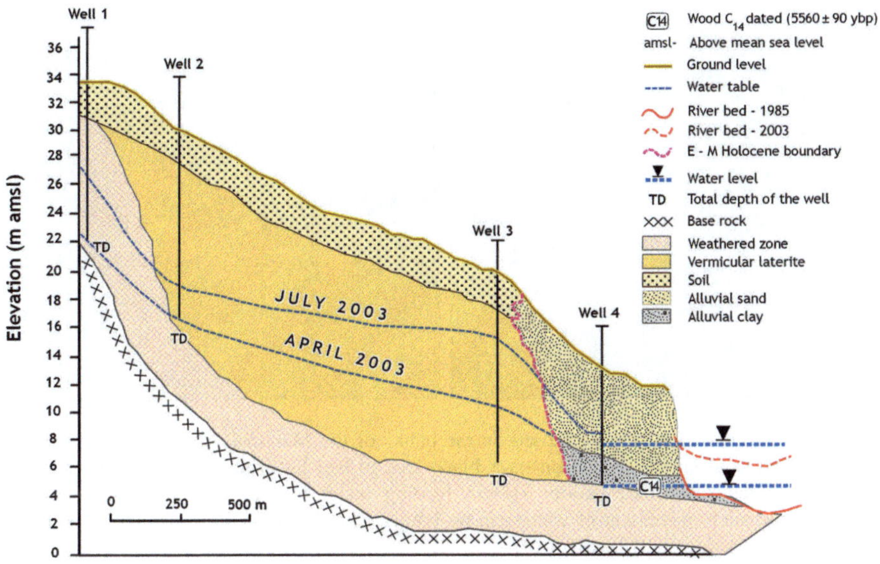

Fig. 4.4 Cross section depicting the connectivity between the water supply wells and the riverbeds (Prasad and Nair 2004; Sreebha 2008)

profound role in the water availability of wells that are close to the river channels and also constructed on older alluvium with sand–clay intercalations (Plate 4.4).

4.5 Changes in Biological Environment

Human activities disrupt the natural flow patterns and ecological processes of rivers with adverse effects on their biological wealth (Petts 1984; Moyle and Leidy 1992). General forms of river degradation due to increasing human population includes physical conversion of natural habitats, water pollution, development of new freshwater resources, introduction of exotic species, and over fishing. Yet, rivers support an extraordinary array of species, many of which are under threat due to habitat destruction. Sand mining has many deleterious direct and indirect effects on the physical, chemical, and biological environments of river systems. However, relatively little attention has so far been made to unravel the effects of sand mining on the riverine biota. Only a few studies are available on the composition of benthic and fish communities affected by changes in sediment texture and habitat loss arising from selective removal of fine aggregates from the riverbed deposits. Indiscriminate sand mining from active channels of rivers cause many adverse effects on the benthic fauna, which inhabits the bottom sandy substratum (Fig. 4.5). Excessive sand extraction from rivers affect the eco-biology of many terrestrial insects whose initial life history begins in aquatic environments

Plate 4.4 Sand and clay layers exposed on the banks of the Manimala river, SW India at two locations. The sand layer exposed on river banks due to bed lowering (channel incision) were once part of the subsurface aquifer system feeding water to the wells in the floodplains. Unfortunately, such movement of water can no longer takes place as the water level has fallen below the permeable (sand) layer

(Plate 4.5). Studies by Sunilkumar (2002) revealed that many aquatic organisms, especially the benthos, are affected severely by clandestine sand mining. The organisms identified in the unaffected reaches of the river include different species of mayfly, dragonfly, chironomids, caddisfly, and other insects of the order Diptera. Interestingly, there is a drastic decrease in the population of dragonfly in the sand mining-hit areas in recent years. The dwindling numerical abundance of benthic fauna might be attributed to indiscriminate sand mining, an activity responsible for the destruction of habitats for their early development in the life cycle. Dragonfly is a beneficial insect to mankind as it is a mosquito predator. But it is unfortunate that this biological control of mosquitoes is decreasing due to destruction of dragonfly nymphs during the sand extraction processes, before their emergence from riverine habitats in adult forms. Sand mining can also negatively affect the survival and dispersal of benthic organisms belonging to the groups Polychaeta, Crustacea, and Mollusca. Dispersal of eggs and larvae is an important aspect of the biological processes of aquatic organisms. From the fisheries point of view, loss of food in the form of benthic invertebrates is a major negative impact, which will ultimately lead to the decline of inland fishery resources.

A few studies on biodiversity, ecological structure and functional processes of fishes are highlighted here to reveal the role of sediment characteristics and habitat alteration on fishes population in aquatic ecosystems, especially rivers. The first and foremost reason for using fishes in riverine biodiversity monitoring is that they are enormously diverse, with different species reflecting different environmental conditions (Moyle and Cech 1996). Further, fishes often have a major bearing on the distribution and abundance of other organisms in the waters they inhabit. Another significant point to highlight is that fishes form an important ecological link between aquatic and terrestrial environments as they form food for many terrestrial

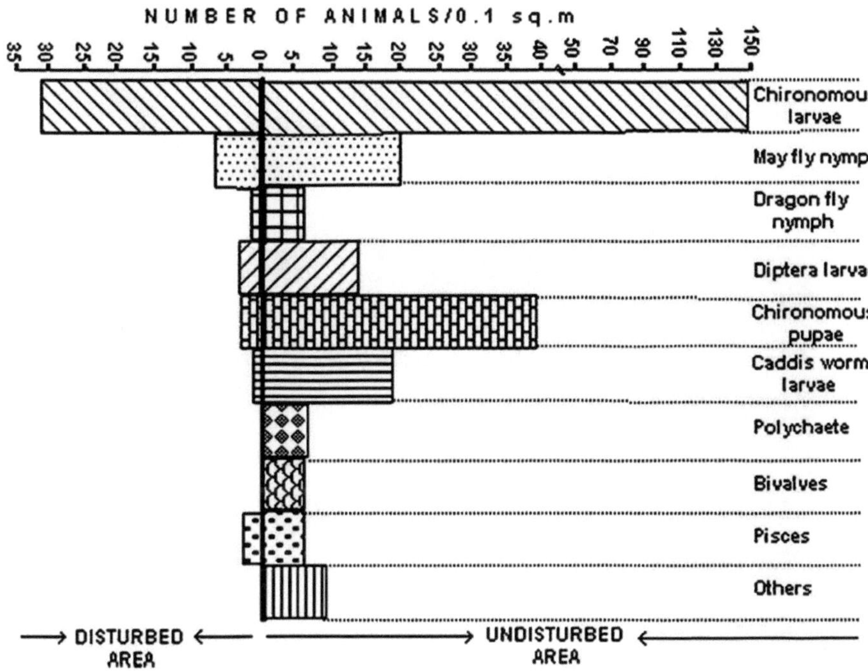

Fig. 4.5 Average abundance (0.1 m^2 area) of benthic organisms in the Achankovil river (Sunilkumar 2002)

organisms, including aves and reptiles. Most of the fishes are open substrate spawners that do not guard their eggs (FAO 1998). The spawning substrata of fishes can be grouped into five categories—(a) Pelagophils: Eggs are nonadhesive, naturally buoyant, and develop while being carried by the water current. Larvae are strongly phototropic and swim actively; (b) Lithophils: Eggs stick to stones and gravel; (c) Phytolithophils: Eggs stick to submerged plants but other substrata are utilized in their absence; (d) Phytophils: Eggs adhere to submerged macrophytes and (e) Psammophils: Eggs are laid on sand or fine roots associated with sand, washed by running water. According to Freeman et al. (1997), fishes preferably occupy areas that best support survival, growth, or reproduction.

The riparian habitats comprise a transition between the river and the upland portion of the watershed. Vegetation that grows at the interface of the river and adjacent riparian habitat and shades the water is known as Shaded Riverine Aquatic habitat (SRA). The SRA vegetation is composed of overhead cover and instream cover. The overhead cover comprises overhanging riparian vegetation and the instream cover comprises woody debris, such as roots and trunks, and aquatic plants. Sand mining has deleterious effects on the riparian vegetation spread along the banks of the rivers. The riparian canopy regulates stream water temperature through shadowing and provides organic matter via litter fall, while

Plate 4.5 Clandestine sand mining activities from rivers affect the habitat of terrestrial insects whose initial life history begins in aquatic environments (*Source* Sunilkumar 2002)

their root systems stabilize the bank and filter lateral sediment and nutrient inputs, thereby controlling stream sediment and nutrient dynamics (Naiman and Decamps 1997). In short, the destruction of riparian vegetation is a real threat to the river environment (Plate 4.6).

Aquatic vegetation plays an important role in maintaining and improving the health of the river environment. However, these are more likely to be of greatest significance for fisheries when a water course has little or no variation in physical structure. The surfaces of submerged leaves are sites of primary and secondary production of microalgae and bacteria, which can rival that of phytoplankton and bacterophils in water column. The community serves as food for grazing invertebrates and protozoa, contributes to bio-purification of organically polluted water courses, and can be a substantial source of planktonic microorganisms (Murphy 1998). Nutrient recycling is an important function accomplished by instream vegetation. They act as the physical link between water and air for many invertebrates, e.g., caddis, which are food for fishes, have aquatic larval stages and

Plate 4.6 Human interventions and state of riparian vegetation. **a** Mechanized clearing of riparian vegetation; **b** Sand mining from the riparian areas is a threat to the natural vegetation

aerial adults. TSS introduced in the water column by sand mining can have profound effect on instream vegetation (Plate 4.7). Turbidity due to the carriage of silt in the water mass may physically block out light penetration through the water column (Sharip et al, 2014), thus reducing the activity of photosynthesis and thereby lowering the levels of primary production.

Plate 4.7 Mud water rivers! Many small rivers in the southwest coast of India (at their storage zones) became mud water rivers because of high load of particulate matters derived from sand mining close to river banks (**a**) and washing of clayey sand (low grade sand) for improving their quality (**b**)

4.6 Changes in Socioeconomic Environment

Social effects are influenced significantly by environmental, social, and economic factors. However, it should be emphasized that quantifying this socio-economic effect is a difficult task. On the other hand, an assessment of socio-economic impacts of river sand mining would be helpful in wise decision-making regarding river management. Though some subcomponents of sand mining may improve the social condition (e.g., generation of income, local revenue, employment, etc.), a majority of the activities will have negative effects on the environment as well as on the society in the long run. Those who are adversely affected by the mining activities include the riparian land owners/residents.

The common arguments favoring river sand mining is that the mining activity creates positive socioeconomic impacts in the area through employment and revenue generations. River sand mining can potentially provide a significant source of revenue through profit-related royalty payments and through fixed taxation (Waelde 1992). The most significant incremental socioeconomic impact will relate to some small employment generation locally and further a field for drivers and suppliers of goods and services. Besides, it should also be noted that the potential for vehicle movement could degenerate roads and highways. This cost is arguably borne by the vehicle operators in their additional road tax. Nevertheless, it is unlikely that this revenue will be fed back to the mining areas where their impact is greatest.

The drastic depletion of the groundwater table due to excessive removal of river sand had already made many rural water supply schemes defunct. It was found that indiscriminate sand mining close to the bridges and filtration tanks of many drinking water schemes is continuing without much interruption even today. Water scarcity during dry periods have become a serious issue in many river stretches of the Vembanad lake catchments (Plate 4.8). The costs of repairing or replacing

Plate 4.8 Impact of sand mining on rural water supply schemes: **a** A water intake structure is exposed due to river bed lowering; **b** Direct pumping of water into the newly constructed intake pit; **c** A check dam constructed for raising water level in the river reach; **d** Tilting of the water intake due to uncontrolled sand mining; **e** and **f** The lowered river channel and the newly constructed water intake structure for the rural water supply scheme

structures damaged by riverbed lowering from sand mining activities will be huge and far higher than the revenue generated from mining activities. Impact on human health is also typically mediated through physical impacts of river sand mining. For example, changes in water flow and water quality may increase the prevalence of communicable diseases. Deterioration of the quality of water used for drinking, bathing, and other domestic purposes is a serious problem in the study area. These effects can ultimately result in permanent and irreversible ecological transformations, rendering mined stretches of rivers useless for subsequent development or use. Parker (1996) states that any adverse effects on the hydrological environment of developing countries will tend to have a corresponding effect on the health of local communities. In many parts of the developing world, communities near extractive operations depend on untreated surface and groundwater for their drinking and other domestic water requirements. Extraction from riverbanks and beds and the resultant generation of particulates, chemical pollutants such as diesel in the water, therefore pose a particular health risk to workers involved in sand mining. Drowning incidents of livestock are also common in midland and lowland areas, but no account exists to highlight the loss of life of livestock and other pet animals. The rivers in the mining-hit areas have turned into death traps due to pit excavation and migration of the pits upstream and downstream during high flow regimes.

Besides the river sand mining's potential for employment generation that is particularly important with regard to livelihood of workers engaged in sand mining, the process is also related to the livelihood issue of local people in the form of resource availability. Effects of sand mining on resource availability include: (a) Loss of access to clean water—used for drinking, bathing, cleaning, irrigation, etc., (b) Loss of land and access to land, (c) Reduced access to food, and d) Loss of trees and vegetation. Loss of productive agricultural lands in the physiographic provinces like highland, midland and lowland, loss of livestock, etc., can have adverse effects on livelihood issues of inhabitants of the river basin (Plate 4.9). Impacts of sand mining on sediment characteristics and water quality can also have marked effects on the levels of fish stock, ultimately disturbing the local food security and protein availability of the people. The fishermen communities in areas close to river basins were in difficulty due to drastic decline in fish catch caused by sand mining over the years. Most of them are traditionally engaged in fishing, but indiscriminate scooping of river sand—the medium that offers the breeding and spawning grounds for many economically important fish species, is one of the major drivers for the observed decline. Conflict between the miners and fishermen communities is frequent in sand mining hot spots where bank failure incidents are also frequent.

It is now well established that extensive sand mining results in habitat destruction of aquatic organisms. The decline of benthic fauna consequent to indiscriminate sand mining can adversely affect survival of carnivorous and omnivorous fishes. At the same time, it also forms food for a group of amphibians and terrestrial. Therefore, these insects constitute an important component of the food chain of both aquatic and terrestrial environments, thereby playing an important role in maintaining the ecological balance and energy flow between these systems.

Benthic fauna of river systems are biological indicators of an area. Occurrence and distribution of these organisms not only indicate good ecological conditions, but also the water quality of the river system. Examination of trends in freshwater fishes studied from different parts of the world points to the fact that most of fauna

Plate 4.9 Riverine fishery is one of the major contributors to local economy in many small rivers of southwest India. **a** A family engaged in riverine fishery; **b** A fish landing center; **c** and **d** A few scenes from fish collection process; **e** and **f** Collected fishes are ready for sale

are markedly declining, indicating the need for immediate conservation measures. Habitat destruction of natural spawning and breeding grounds of fishes through sand extraction and construction of physical obstructions across rivers have resulted in the drastic decline in the population of fishes. Information regarding migration, breeding behavior, and spawning grounds of threatened fishes should be generated and gathered through extensive field investigations and studies. Aquatic and riparian vegetation is a prime factor in determining the value of riverine fishery and its potential. The direct scooping of sand from active channels and overbank areas can have major negative impacts. Efforts should be made to control the various human interventions in freshwater environments.

4.7 Vegetation Effects

Uncontrolled sand mining removes, disturbs, and/or destroys vegetation in the bars and riparian areas. This could increase the water temperature (Beschta et al. 1987; Sullivan et al. 1990). Removal of standing trees and drowned vegetations in the bars reduces the river load of large woody debris, which play an important role in creating habitats and supplying materials (Bisson et al. 1987; Murphy and Meehan 1991).

4.8 Damage to Infrastructure Facilities

Uncontrolled sand mining can cause damage to infrastructure facilities attached to river channels. An ancient dam constructed about 1,500 years ago across Dedura Oya in Sri Lanka is almost completely damaged due to indiscriminate sand mining. Further, irrigation intake structures such as barrages and pump houses along rivers have been totally disrupted due to indiscriminate sand mining and subsequent channel incision. In Kerala in southwest India, many intake structures and lift irrigation schemes constructed in Pampa and Periyar rivers were abandoned due to riverbed lowering and damage caused to these structures. In addition to this, many bridges and highways are also affected adversely due to indiscriminate sand mining.

4.9 Flood Plain Pits Capturing

Indiscriminate wet pit mining of sand from floodplain areas of rivers close to active channels and subsequent enlargement of the pits due to continued sand mining leads to drastic reduction in the buffer area separating the floodplain pit from the active channel. During high-flow regimes, the alluvial wall separating the floodplain pit and the active channel collapses as the river captures the floodplain pit (Fig. 4.6). Capture of several pits in a given floodplain region leads to channel widening and disfiguration of natural river course (Fig. 4.7). Plate 4.10 shows field

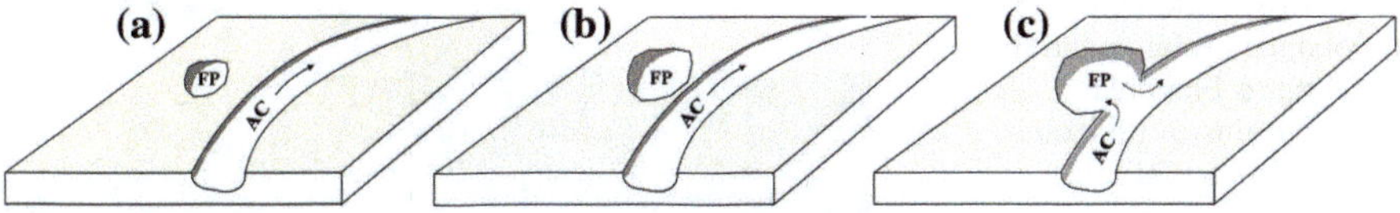

Fig. 4.6 Schematic presentation of stages of pit capturing; **a** Binging of pit excavation for construction grade sand; **b** Reduction in the buffer area between the floodplain pit (FP) and active channel (AC) due to indiscriminate wet pit mining and **c** Pit capturing under high flow regime. Arrow indicates direction of river flow

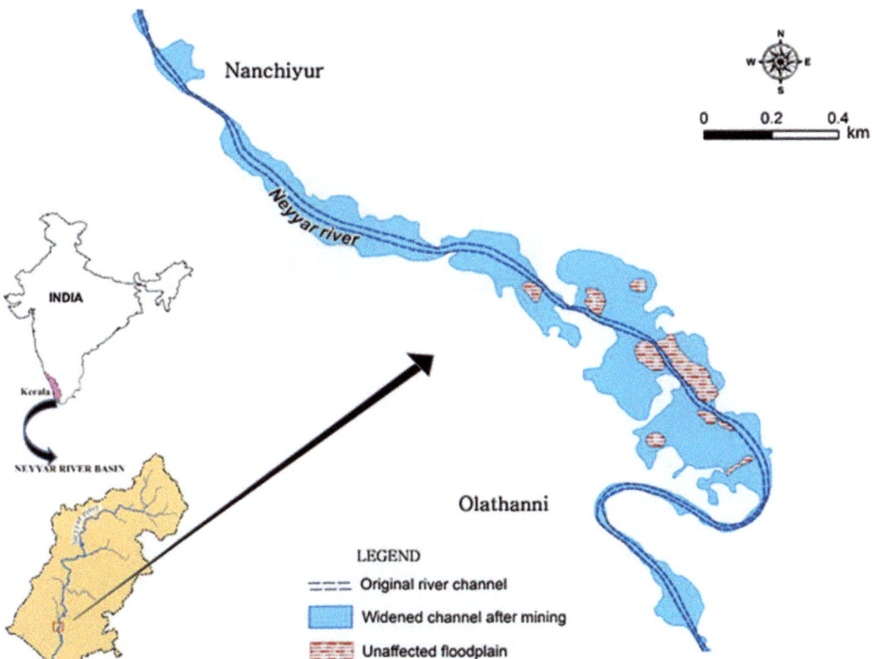

Fig. 4.7 River widening due to floodplain mining for construction grade sand from Neyyar river, southwest India

evidences in the different stages of pit capturing. Pit capturing alters the floodplain hydrology and sediment budget of the river in its storage zone.

Plate 4.10 Field evidences of floodplain mining and pit capturing. **a** Floodplain mining begins in Ithikkara river; **b** The buffer area separating river and the floodplain pit reduces to dangerous level (Kabani river); **c** River captures a floodplain pit (Neyyar river)

4.10 Impacts of River Sand Mining on Coastal Marine Environment

Sand is one of the world's most plentiful resources on the Earth's crust. Sand and gravel that form aggregates for construction, account for the largest material volumes mined in the world. The global production in the year 2000 was estimated to exceed 15 billion tons. Sediment is supplied to coastal environment through different processes—headland erosion, river transport, and offshore sources (Kondolf 1997). Although rivers bring a significant quantity of sediment to the coast, only a small percentage of it is deposited in the nearshore environment. Generally, coarse sand, gravel, and other larger particles are detained near the base of the eroding surfaces, the finer sediments are deposited in the floodplain, bays or lagoons, and at the shoreline as delta deposits. The very fine silt and clay fractions that typically make up a significant portion of the eroded material move offshore where it eventually settles. The sand fractions deposited at the shoreline is gradually moved along the coast by waves and currents, and provides nourishment for beaches. Eventually, a substantial proportion of this littoral material is often lost to offshore areas. The sediment yield from upstream stretches increases with increase in relief of the drainage basin but decreases with size, which explains the importance of the small rivers in coastal sediment transport, as is the case of Kerala State. Other factors that determine the sediment load are climate, rainfall, and geological conditions of the area.

Reduced sediment supply in the coastal environments through reduction in sediment delivery from rivers and streams results in creation of undernourished beaches or shrinkage of beaches accelerating both beach and cliff erosion (Inman 1985; Kondolf 1997). Beaches are the natural barriers protecting land areas from coastal erosion. Sediment transport from the headland that replenishes the beaches plays a pivotal role in reducing the fury of the waves in high-energy regimes, as it reduces the hungry water effect, i.e., the destructive effect of sand-free waters. It is seen that a substantial quantity of sand from beaches is also mined illegally for various purposes including building constructions. Beach sand removal is prohibited in many parts of the world, as the activity along with reduced river sediment supply from upstream stretches exacerbate erosion leading to higher cost for coastal protection. Reduced sediment transport from river systems also affects directly the beach sand replenishment in coastal areas, which minimizes the available land surface since it accelerates coastal erosion (Elisebeth et al. 2010; Allenbach 1999).

The reduced supply of sediment from upstream sources interrupts the natural development of dunes and other landforms in the marine environment. The normal topography and coastal geologic settings will be altered due to the accelerated scouring of beaches and reduced deposition of sand. One of the most significant morphological impacts of rampant river sand extraction is on the backwater systems generally running parallel to the shoreline, locally known as *kayals*. Their formation is linked with subsidence/lowering of sea level and emergence of sand bars, which is interrupted by the rampant scooping of sand from the upstream areas. In addition, there are several reports of aggravated salt water ingression consequent to mining-induced channel incision, especially in areas close to river mouth zones, which accentuates the water crisis situation further.

References

Allenbach M (1999) Mission d'expertise sur l'érosion du lineare côtier de l'île de Wallis—Evolution naturelle eta nthropisation, gestion et aménagement littoral Laboratoire des Géosciences, Université de la Nouvelle-Calédonie

Anooja S, Baijulal B, Maya K, Sreebha S, Padmalal D (2011) Impact of sand mining on river bed changes and bed material characteristics—a case analysis. National seminar on mining of river sand and its impacts on the environment, CWRDM, Kozhikode, pp 173–181

Babu KN, Sreebha S (2004) Evaluation of nutrient budget of the rivers and adjoining backwater-nearshore systems of Kerala (Unpublished report), Centre for Earth Science Studies, Thiruvananthapuram, India, p 118

Beschta Rl, Bilby RE, Brown GW, Holtby LB, Hofstra TD (1987) Stream temperature and aquatic habitat: fisheries and forestry interactions. In: Salo EO, Cundy TW (eds) Stream-side management: fishery and forestry interactions, Institute of forest resources, University of Washington, Seatle, Washington, pp 191–232

Bisson PA, Sullivan K, Neilson JL (1987) Channel hydraulics, habitat lose and body form of juvenile coho Salmon, Steelhead trout and cutthroat trout. Trans Am Fish Soc 117:262–273

Bull WB, Scott (1974) Impact of mining gravel from urban stream beds in the southwestern United States. J Geol 2:171–174

Collins B, Dunne T (1989) Gravel transport, gravel harvesting, and channel-bed degradation in rivers draining the Southern Olympic Mountains, Washington, USA. Environ Geol Water Sci 13:213–224

Collins BD, Dunne T (1990) Fluvial geomorphology and river-gravel mining: a guide for planners, case studies included. California Department of Conservation, Division of Mines and Geology Special Publication, California, vol 98, p 29

Elisabeth W, Allenbach M, Hermann H (2010) Climate change and the acceptance of coastal sand mining on Wallis Island. PPME—EA 3325, Université de la Nouvelle-Calédonie, BP R4, Nouméa, Nouvelle Calédonie

Erskine WD (2008) Channel incision and sand compartmentalization in an Australian sand stone drainage basin subject to high flood variability. In: Proceedings of a symposium held in Christchurch, Newzealand, Dec 2008, Sediment dynamics in changing environments, IAHS publication, p 1–8

Evor B, Holland M (1989) Surface and groundwater management in surface mined—land reclamation. California Department of Conservation, Division of Mines and Geology, Special Report 163, p 39

FAO (1998) Rehabilitation of rivers for fish, food and agriculture. United Nations Organization. Fishing News Books, Oxford, p 260

Freeman MC, Bowen ZH, Crance JH (1997) Transferability of habitat suitability criteria for fishes in warm water streams. North Am J Fish Manage 17:20–31

Goodwin P, Florsheim J, Rubin Y (1992) Sonoma County Planning Department, Hydrologic Aspects, Aggregate Resource Management Plan Update and EIR (Unpublished report), Philip Williams and Associates, Ltd, p 71

Hartfield P (1993) Head cuts and their effects on freshwater mussels. In: Proceedings UMRCC symposium, Upper Mississippi river conservation committee, Illinois, p 131–141

Heede BH (1986) Designing for dynamic equilibrium in streams. Water Resour Bull 22:351–357

Inman DL (1985) Budget of sand in southern California; river discharge vs cliff erosion. In: Proceedings from a conference on coastal erosion, California Coastal Commission, pp 10–15

Kanehl P, Lyons J (1992) Impacts of instream sand and gravel mining on stream habitat and fish communities including a survey on the Big Rib river, Marathon County, Wisconsin. Wisconsin Department of Natural Resources, Research Report 155, p 58

Knighton D (1984) Fluvial forms and processes. Edward Arnold/Hodder and Stoughton, London, p 320

Kondolf GM (1997) Hungry water: effects of dams and gravel mining on river channels. Environ Manage 21:533–551

Kondolf GM (1998) Large scale extraction of alluvial deposits from rivers in California: geomorphic effects and regulatory strategies. In: Klingeman PC, Beschta RL, Komar PD, Bradley JB (eds) Gravel bed rivers in the environment. Water Resources Publications, Colorado, pp 455–470

Kondolf GM, Matthews WVG (1993) Management of coarse sediment in regulated rivers of California. University of California Water Resources Centre, California, p 40

Kondolf GM, Smeltzer M, Kimball L (2002) Freshwater gravel mining and dredging issues. Report prepared for Washington department of fish and wildlife, Washington Department Of Ecology and Washington Department of Transportation, p 122

Lake RG, Hinch SG (1999) Acute effects of suspended sediment angularity on juvenile coho salmon. Can J Fish Aquat Sci 56:862–867

Lake County (1992) Lake County aggregate resource management plan. Draft, Lake County Planning Department, Resource Management Division, California, p 50

Meador MR, Layher AO (1998) Instream sand and gravel mining: environmental issues and regulatory process in the United States. Fisheries 11:6–13

Moyle PB, Cech JJ (1996) Fishes: an introduction to ichthyology. Prentice Hall, New Jersey, p 295

Moyle PB, Leidy RA (1992) Loss of biodiversity in aquatic ecosystems: evidence from fish faunas. In: Fiedler PL, Jain SK (eds) Conservation Biology: the theory and practice of nature conservation, preservation and management. Chapman and Hall, New York, pp 127–169

Murphy ML (1998) Primary production. In: Naiman RJ, Bilby RE (eds) River ecology and management: lessons from the pacific coastal ecoregion. Springer, New York, pp 144–168

Murphy ML, Meehan WR (1991) Stream ecosystems. In: Meehan WR (ed) Influence of forest and rangeland management on salmonid fishes and their habitats, Ame Fish Soc Spl, Publ. No19, Bethesda, Maryland

Naiman RJ, Decamps H (1997) The ecology of the interfaces: riparian zones. Annu Rev Ecol Syst 28:621–658

Nelson KL (1993) Instream sand and gravel mining. In: Bryan CF, Rutherford DA (eds) Impacts on warm water streams: guidelines for evaluation. Southern Division, American Fisheries Society, Arkansas, pp 189–196

Padmalal D (1992) Mineralogy and geochemistry of the sediments of Muvattupuzha river and Central Vembanad estuary, Kerala, India. Ph.D thesis (Unpublished), Cochin University of Science and Technology, Kochi, India

Padmalal D, Maya K, Narendrababu KN, Baijulal B (2010) Environmental appraisal and sand auditing of Manimala river, Kerala, India. Project Completion Report submitted to Revenue Department, Government of Kerala

Parker D (1996) Environmental assessment and auditing of mining operations—an international perspective. In: Proceedings of the IBC UK conferences on the environmental management of mining operations, London

Petts GE (1984) Impounded rivers. Wiley, Chichester, p 326

Prasad RR, Nair KM (2004) Integrated hydrogeological investigations and multipronged water conservation in selected watershed of Achankovil river basin. Report of People's Research Organisation for grass root environmental science service (Progress), Hyderabad, p 43

Sandecki M (1989) Aggregate mining in river systems. Cal Geol 42:88–94

Scott KM (1973) Scour and fill in Tujunga wash: a fan head valley in urban southern California—1969. US Geological Survey Professional Paper, p 70
Sharip Z, Taqiuddin A, Zaki A (2014) The effects of season and sand mining activities on thermal regime and water quality in a large shallow tropical lake. Environ Monit. Accessed doi:10.1007/S10661-014-375-4
Sheeba S, Arun PR (2003) Impact of sand mining on the biological environment of Ithikkara river—an overview. In: Proceedings of 15th Kerala science congress, Thiruvananthapuram, pp 806–807
Starnes LB (1983) Effects of surface mining on aquatic resources in North America. Fisheries, 8: 2–4
Stevens MA, Urbonas B, Tucker LS (1990) Public—private cooperation protects river. APWA reporter, pp 25–27
Sullivan K, Tooley J, Doughty K, Caldwell JE, Knudson P (1990) Evaluation of prediction models and characterisation of stream temperature regimes in Washington. Timber, Fish and Wildlife Programme Report, TFW-WQ3-90-006, p 224
Sunilkumar R (2002) Impact of sand mining on benthic fauna: a case study from Achankovil river—an overview. Catholicate College, Pathanamthitta district, Kerala, p 38
Thomas RB (1985) Estimating total suspended sediment yield with probability sampling. Wat Resour Res 21:1381–1388
Thrivikramaji KP (1986) River metamorphosis due to human intervention in Neyyar river basin. Final Technical Report, Department of Environment, Government of India, p 112
UNEP (1990) Environmental guidelines for sand and gravel extraction projects. Environmental guidelines, No. 20, United Nations environment programme, Nairobi, p 37
Wahler Associates (1981) Cache Creek aggregate resources, Yolo County California. Unpublished report to Yolo County
Whiting PJ (1998) Floodplain maintenance flows. Rivers 6:160–170
Williams GP, Wolman MG (1984) Downstream effects of dams on alluvial rivers. US Geol Surv Prof Pap 1286:89
Yang CT (2003) Sediment transport: theory and Practice. Mc.Graw Hill, New York, p 396

Chapter 5
Sand Mining: The World Scenario

Abstract River sand mining is widespread in most of the developing and developed countries. Of the different types of fluvial systems, small rivers (catchment area <10,000 km^2) are the worst affected due to indiscriminate sand mining than large rivers as the area available for dissipation of negative externalities is low in small river basins. This chapter deals with a brief account of the environmental effects of river sand mining reported from different parts of the world like Arizona, Italy, Kenya, Jamaica and Costa Rica, Malaysia, China, Sri Lanka, Nepal, Maldives and India.

Keywords River sand mining · Head cutting · Salt water ingression · Small rivers of Western Ghats

5.1 Introduction

Impacts of sand and gravel mining from various depositional environments including the river ecosystems have been reported from many parts of the world. However, lack of adequate scientific studies on the adverse impacts of sand mining on various environmental components of river and related ecosystems in peer reviewed journals is a major lacuna challenging regulatory efforts and awareness creation at various levels. Among the different types of mining, extraction of sand for building constructions is one of the largest industries not only in volume but also in value. This chapter deals with the salient observations on the extraction of construction grade sand reported from different parts of the world.

5.2 Arizona and California

Many cases of river sand and gravel mining induced bridge failure, headcut migration, and bank erosion have been reported in Arizona by different researchers and/or organisations (Bull and Scott 1974). In river systems, bed load transport of

D. Padmalal and K. Maya, *Sand Mining*, Environmental Science and Engineering,
DOI: 10.1007/978-94-017-9144-1_5,

sediments have been adversely affected by human interventions in the river basin like reservoir constructions, sand/gravel extraction from in-channel and off-channel areas, etc. In some instances, the effects of river sand and gravel mining may be difficult to distinguish from the natural processes operating in the system. However, studies of Kondolf and Swanson (1993) could differentiate two types of responses in the Stony creek of California (Fig. 5.1). Incision and channel instability caused by instream gravel mining which is frequent in many rivers of California (Kondolf and Swanson 1993). Figure 5.2 shows the channel incision recorded in the longitudinal profile of the Russian river in California (Kondolf 1997).

Aggregate mining generally occurs within 30–50 miles of the central market because cost of transport is the primary expense in the sand and gravel mining industry (Meador and Layher 1998). In Oregon, the focus of instream aggregate mining operations is confined to cities and major roadways. The market for the aggregate materials includes Portland, Salem, Albany, Eugene, Roseburg, and many other smaller municipalities and counties (Castro and Cluer 2003).

The California Division of Mines and Geology (1987) estimated that the average per capita consumption of sand and gravel was 7.4 tonnes per person per year. An average child in California requires a lifetime supply of 620 tonnes of sand and gravel. As the cost of sand and gravel largely depends on haulage distance, it doubles every 40–50 miles of transport by truck (CDMG 1994). The sand mining industry is responsible for causing damages to bridges, siphons and other river associated engineering structures. It is now well understood that indiscriminate sand and gravel mining not only aggravate the pace of river degradation but also undermine foundation structures of the engineering constructions. Instream mining may also cause damage to fishes and wildlife resources by promoting degradation and undesirable changes in stream morphology.

Extraction of instream gravel sources causes damage to riparian vegetation, groundwater resource, water quality, fish, and wildlife. Human interventions on the watershed include activities like timber/firewood extraction, livestock grazing, fires, agriculture, road building, and urbanization. Alterations of the stream channel include riparian vegetation removal, channelization, sand and gravel mining, dams and check dams, water diversions, levees, and bridges. Rivers respond to these processes with changes in bank erosion, aggradation, flooding, channel straightening, etc. The stream may change its grade, sinuosity, depth, width, sediment transport, and radius of curvature in response to these human-induced changes (CDWR 1994). Three general types of aggregate resources are available in California: (a) instream alluvial sands, (b) offstream alluvial sands, and (c) hard rock based products. In a study carried out in Southern Missouri, Suzanne (2002) revealed that indiscriminate extraction of sand and gravel alters the sediment budget of the river leading to channel instability, increased turbidity, degradation of habitats, etc. Therefore, rock-based products (i.e., manufactured sand) are used in greater quantities than instream sand in many parts of Southern Missouri.

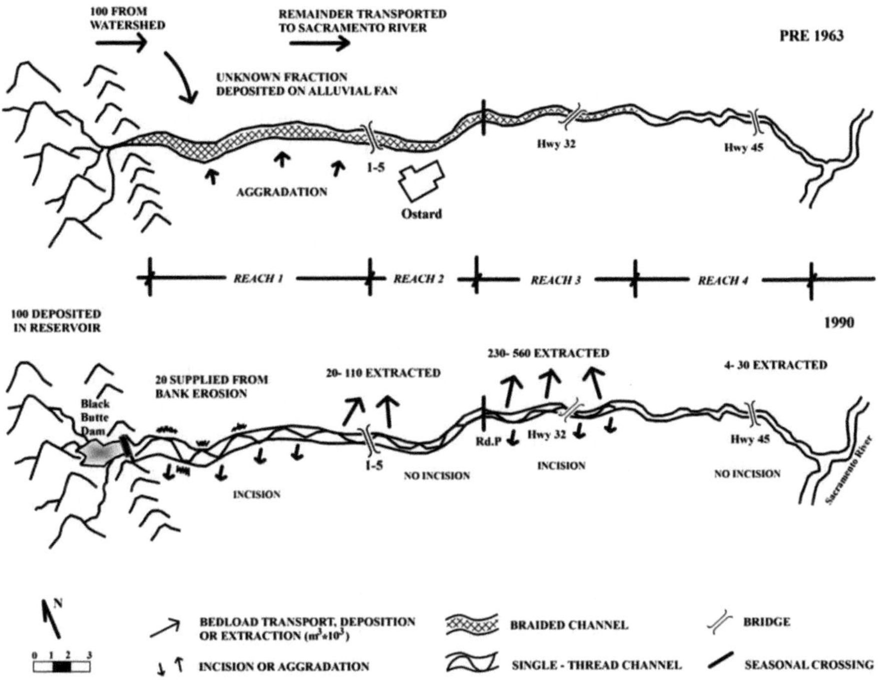

Fig. 5.1 Illustration of bed load sediment budget for Lower Stony Creek prior to construction and after construction of Black Butte Dam. All values of gravel flux are in $m^3 \times 10^3$ (*Source* Kondolf and Swanson 1993)

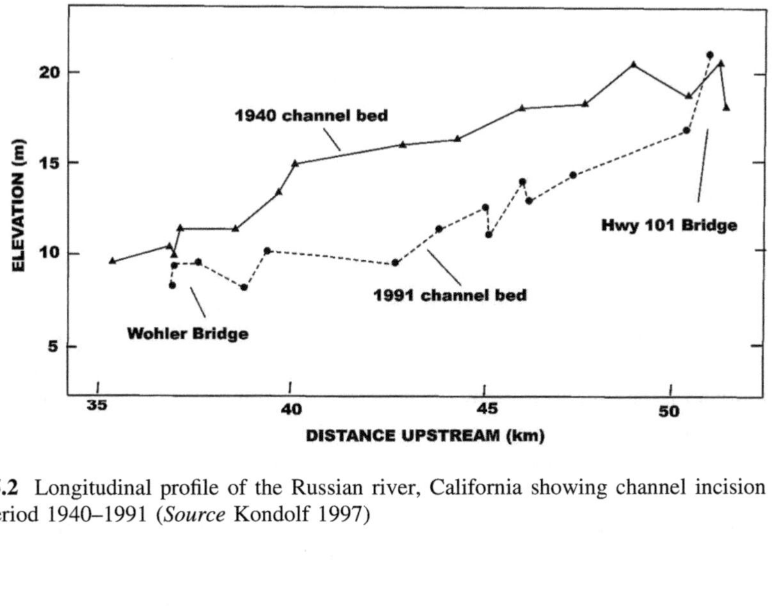

Fig. 5.2 Longitudinal profile of the Russian river, California showing channel incision during the period 1940–1991 (*Source* Kondolf 1997)

5.3 Italy

The terrain areas of Italy are drained mainly by small rivers whose drainage area varies between 1,000 km^2 and 10,000 km^2. Only three rivers in Italy have drainage area greater than 10,000 km^2. They are Po (70,091 km^2/651 km), Adige (11,954 km^2/410 km), and Tevere (17,556 km^2/396 km). The studies of Rinaldi (2003). Surian and Rinaldi (2003) revealed that many Italian rivers were degraded considerably due to human interventions in the last century. The most common morphological changes noticed in the Italian rivers are bed-lowering, channel narrowing and changes in channel pattern. Channel incision of 3–4 m is very common in the storage zones of many rivers. In certain rivers draining through Emilia, Romagna, Marche, Abruzzo, and Calabria regions, the channels have been incised up to 10 m or even more at certain locations. Table 5.1 shows a comparative evaluation of the channel adjustments of some of the important Italian rivers with that of the rivers of Australia, California, China, Sri Lanka, and India. Human interventions like construction of dams, sand and gravel mining, channelization, land use changes, etc., have been responsible for the observed changes in river bed morphology. Among these, sand mining is perhaps the most detrimental factor leading to river degradation (Surian and Rinaldi 2003). The activity was very intense in the period between 1950s and 1970s, although the process has continued at a slower pace later on (Surian and Rinaldi 2003). For example in the Po river, instream sand mining increased from about 3 million m^3/year to about 12 million m^3/year during the period 1960–1980, and later reversed to the pre-1960s scenario because of stringent regulation of the process by the civic authorities (Lamberti 1993). Rinaldi et al. (2005) reported that sediment deficit caused by instream mining is the major cause of channel incision in many of the Italian rivers. Channel incision alters significantly the frequency of floodplain inundation along the river courses, lowers water tables on the river bank areas, damages bridges and leads to loss or impoverishment of aquatic and riparian habitats (Surian et al. 2009).

5.4 Kenya

Like many other developing countries, indiscriminate river sand and gravel mining is a major environmental concern in Kenya as the activity causes irreparable damages to the Kenyan rivers. The commercial mining of river sand was noticed since early 1950's (Baker 1954). Sand mining has grown to be an integral component and a major source of the Kenyan economy with about 90 % of annual sand generation coming from river sources alone. Figure 5.3 shows the sand mining locations in various tributaries of the Athi river, draining through the south eastern part of Nairobi in Kenya. Table 5.2 shows the various on-site and off-site impacts of river sand and gravel mining along with the management strategies recommended to contain the adversities of sand mining operations. According to experts,

Table 5.1 Channel adjustments in rivers and relative causes and effects

Sl. no.	River/Country	Causes	Effects	Reference
1	Brental river (Italy)	Gravel mining; damming	Failure of bridges; loss of groundwater resources	Castiglioni and Pellegrini (1981a, b)
2	Po river (Italy)	Gravel and sand mining; River engineering; human interferences	Undermining of bank-protection structures; loss of agricultural land; loss of groundwater resources	Surian and Rinaldi (2003)
3	Hunter river (Australia)	Damming; sand and gravel mining; annual sand extraction of about 0.2 million tonnes downstream of Glenbawn dam	Channel incision; bank erosion	Erskine et al. (1985)
4	Frasier creek (California)	Sand and gravel mining	Channel incision	Kondolf and Swanson (1993)
5	Pearl river (China)	Sand and gravel mining	Channel incision	Lu et al. (2007)
6	Nilwala river(Sri Lanka)	Sand and gravel mining	Channel incision; salt water intrusion in downstream areas; river bank erosion.	Ranjana (2011)
7	Manimala river Kerala, India	Sand and gravel mining	Channel incision; river bank erosion; undermining of engineering structures	Padmalal (2010)
8	Periyar river Kerala, India	Sand and gravel mining; damming	Channel incision; river bank erosion; water table lowering	Padmalal et al. (2011)

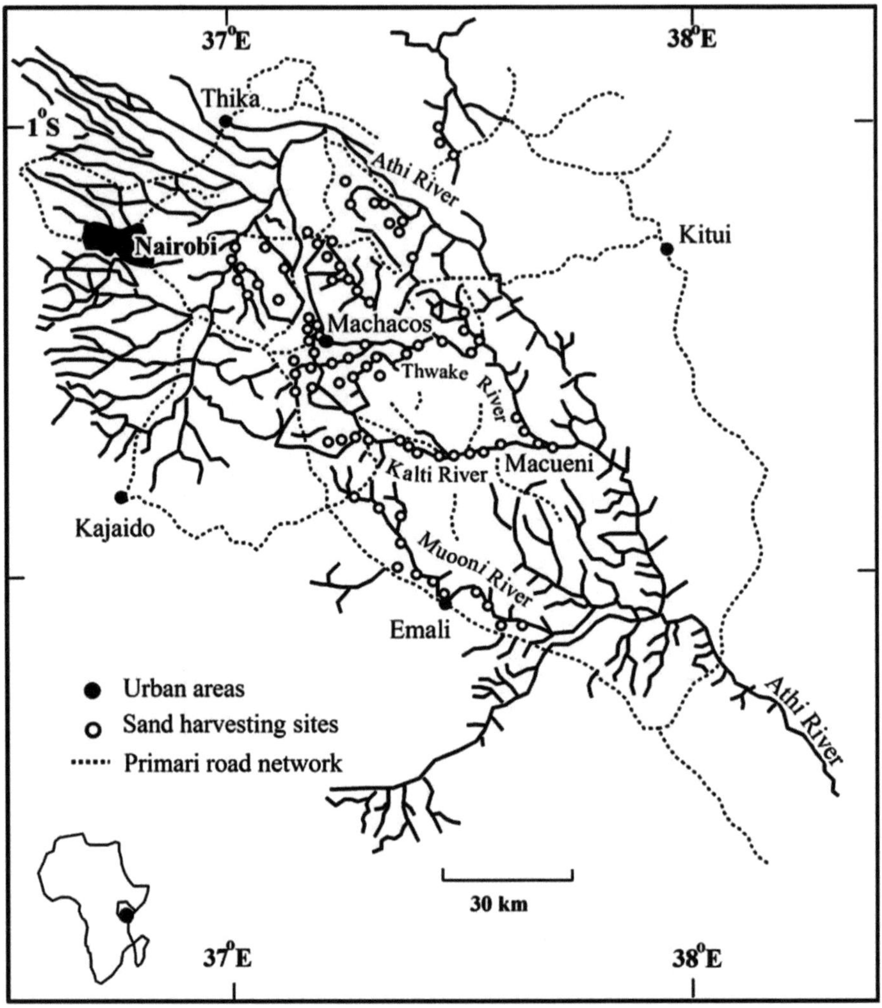

Fig. 5.3 Locations of sand mining sites in South–East of Nairobi, Kenya (*Source* Kitetu and Rowan 1997)

the environmental impacts of river sand mining are not always obvious and hence have long been underestimated (Kitetu and Rowan 1997). But, the Kenyan government has now realized the impact of sand and gravel mining on various components of the aquatic ecosystems through systematic evaluation of the state of its environment (Diang'a 1992). Environmental degradation often led to conflict between miners and local communities who are depended on rural water supply schemes attached to river systems. Considering the adverse impact of river sand and gravel mining, Kitetu and Rowan (1997) made a systematic study for assessing the cumulative long-term environmental effects of the Kenyan river sand industry and its relationship with the under development process.

Table 5.2 Environmental management issues in the Kenyan river sand industry, and other agencies (*Source* Kitetu and Rowan 1997)

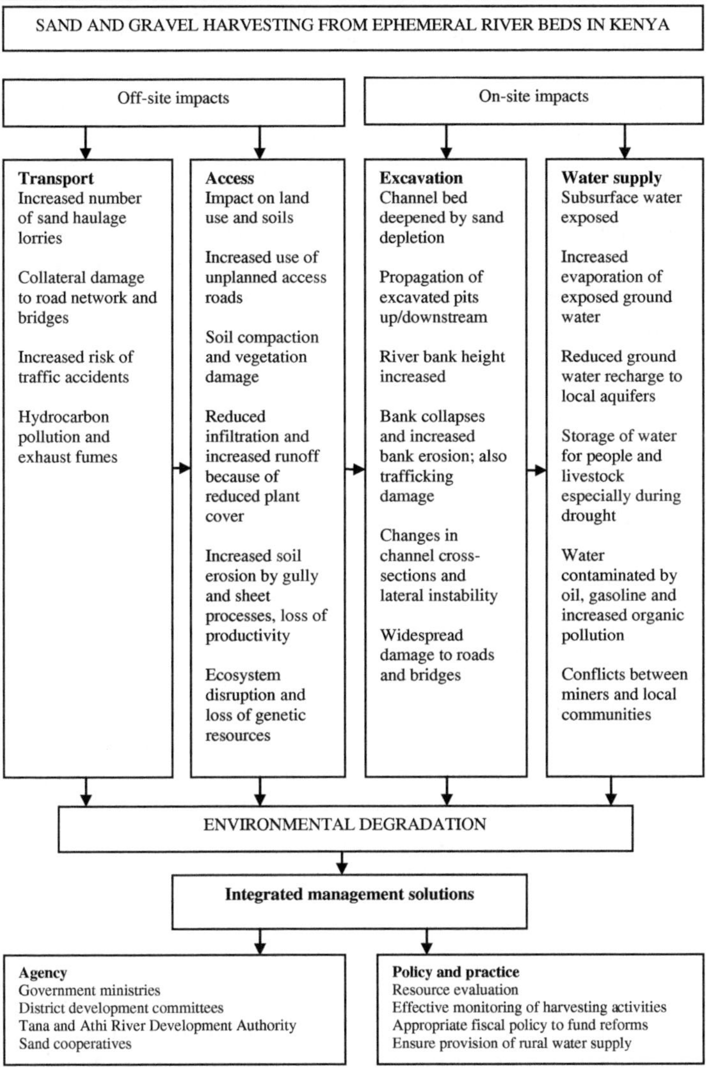

5.5 Jamaica and Costa Rica

In Jamaica and Costa Rica, substantial quantity of sand and gravel occur within the floodplain and terrace deposits and beneath the agricultural lands of the major river basins (Alvarado-Villalon et al. 2003; Farrant et al. 2003). However, sand and gravel extraction is confined mainly to river channel deposits. There are many rivers which can supply sand and gravel, particularly in the Atlantic region. Distribution of the resources is mainly through roadways but a marginal portion of sand is being transported to other places by railroad networks as well. Alluvial sand and gravel extractions in Costa Rica are carried out by numerous small companies using unskilled or semi-skilled labour. As the environmental effects of alluvial sand extraction has been of concern in Costa Rica and Jamaica, investigations have been taken up to assess the impacts of sand and gravel mining and its effects on the environmental components of the rivers, for laying down strategies for environment-friendly mining. In Jamaica, large amounts of sand and gravel have been extracted from the Rio Minho valley and Yallas fan-delta. The total production amounts to about 10 million tonnes per year. Many parts of the active channel have extensively been mined for sand and gravel for civil constructions and maintenance works.

Headward erosion and channel incision are widespread in many of the rivers of Costa Rica and Jamaica.

5.6 Malaysia

In Malaysia, river sand and river stone are extracted widely from the fluvial channels for the production of coarse and fine aggregates. Lack of land-based rock aggregates and its poor quality are some of the factors responsible for the indiscriminate mining of river bed materials. Mining of river sand, gravel and stones are rampant in many of the Malaysian rivers. In Malaysia, river sand and gravel mining requires EIA approval prior to the project commencement. Therefore, the activity is classified under "Prescribed Activity" in the Conservation of Environment order. The Environmental Conservation Department (ECD), Malaysia is responsible for the implementation of the order to ensure that the activity is taking place in an environment-friendly manner. In order to identify the potential impacts of sand, gravel, and stone mining, the ECD, Malaysia has evolved a set of guidelines with a view to ensure that the mining related activities will be carried out with minimum negative impacts (ECD 2000). However, very often mining occurs recklessly which jeopardizes the environmental setting of the river leading to deterioration of its ecosystem structure and functions. Therefore, the Department of Irrigation and Drainage (DID), another statutory body involved in regulation of river sand mining, brought out a set of guidelines namely "River Sand Mining and Management Guidelines'' for enabling the DID engineers and mining operators to get a clear understanding of the theory of sediment transport process that determines the sand replenishment rate and hence limit the volume of sand that can be extracted from the river reach. The application of annual replenishment concept is the key to ensure long-term river channel stability as well as the health of the aquatic and riparian habitats (DID 2009). The study of Ashraf et al. (2011) in Selangor river revealed that on an average 11.73 million ty^{-1} of sand and gravel are being extracted from the active channels and 0.414 million ty^{-1} from the floodplains. The quantity of instream sand mining is reported to be about 40 times higher than the sand replenishment estimated in the gauging stations.

Indiscriminate sand mining has resulted in the lowering of the river bed in the storage zone at a rate of 7--15 cmy^{-1} over the past two decades.

5.7 China

Extensive and large-scale sand and gravel mining since mid 1980s for meeting the ever increasing demand in construction sector has markedly affected the geo-environmental settings of the major river basins of China. The activities reached its severity in the rivers like Pearl, Yangtze etc., (Wang et al. 2012). Illicit sand mining was rampant in and around the fast growing Chinese cities and townships despite stringent control measures and harsh punishments. The wide gap between demand and supply of construction grade sand is a major concern in many parts of China. The demand of construction grade sand in Guangdong's market in 2012 was estimated to be to the tune of 100 million m^3 against an actual supply of just 15 million m^3. Studies revealed that the hydrology, channel morphology and biodiversity of many river systems in China have been irreversibly affected by unabated sand and gravel mining over the past few decades. Uncontrolled sand mining from the Pearl River Delta in South China has altered many of its tributaries. It is estimated that $>8.7 \times 10^8$ m^3 of sand was extracted from the river during the period 1986–2003, which resulted in an average channel incision of 0.59–1.73, 0.34–4.43, and 1.77–6.48 m in the main channel at three major water networking stations in the Pearl river delta. Further, salt water intrusion has reached 10–20 km landward from the level of 1980's (Luo et al. 2007). The ecosystem damage and biodiversity changes are also aggravated due to changes in physico-chemical setting of the fluvial environment resulted from channel incision. The pace of sand mining activities have changed in tandem with the strength of the real estate market in the region. The excavation rate has enhanced manifold during 1990–1995. Almost all the river channels in the region are adversely impacted by the activity. The pace of sand and gravel mining has slowed down due to the Asian financial crisis and subsequent cooling of the real estate market in 1997. Since 2000, the public and government agencies in China have expressed deep concern over the adverse environmental effects of river sand mining and subsequent river bed degradation (Luo et al. 2007). The flood section of the river at Sansin station has experienced only marginal changes from 1984 to 1988, but has been subjected to drastic changes during the post 1988 period. The river channel at this station has been incised to more than 7 m during the period 1988– 2005 indicating the rampant sand extraction from the channel section. The ban on sand mining in rivers like Yangtze has accelerated the use of alternate sand sources. Leeuw et al. (2010) estimated that an amount of 236 million m^3 y^{-1} sand was extracted from the Poyang lake during 2000–2006. This corresponds to 9 % of the total demand of sand in China (Leeuw et al. 2010). Indiscriminate mining of sand from lakes have become a new concern of the people and government agencies as the rivers in the country are being adversely affected by rampant sand mining.

5.8 Sri Lanka

River sources of sand have now become a scarce commodity in Sri Lanka due to over exploitation of sand from the river ecosystems in the past few decades. The demand of construction grade sand has increased manifold after the Tsunami disaster of 2004. The construction industry of Sri Lanka accounts for about 8 % of the country's GDP (Athukorala and Navaratne 2009) and requires about 7–7.5 million cubic meters of aggregates per year (Ranjana et al. 2009). This aggregate requirement is inevitably met from the river beds, floodplains, paleo-channels, and sand dunes. Manual extraction was practiced in Sri Lanka till the 1990s. Mechanical mining for sand became rampant in the country during the post-tsunami period. This has resulted in wide ranging impacts including channel incision and subsequent intrusion of sea water into rivers, collapse of river banks, loss of riparian lands, changes in river morphology, damages to road and other infrastructural facilities, and excessive coastal erosion due to paucity of sand discharge from the hinterland rivers, etc. The situation is rather alarming in the rivers like Kalu Ganga, Walawe Ganga, Kelani Ganga, Nilwala Ganga, Deduru Oya, Ma Oya, and Kirindi Oya.

Environmental problems consequent to river sand mining is severe in the northern part of Vavuniya district (Priyantha and Mayooran 2011). Ranjana et al. (2009) reported that, like the northern part, the rivers in the southern part of Sri Lanka such as Nilwala and Girganga rivers are also affected severely by indiscriminate river sand mining. It has been reported that the mining of sand is extensive in the Nilwala river that drains through the Singherage Natural Forest. Sand mining in this river is carried out upto 45 km from the river mouth located at Bope gode. The river is the only source of drinking water in the Matare district. The first water pumping station in the river, constructed at Nadugala, located about 8 km inland from the river confluence with the sea (i.e., river mouth point) had to be abandoned due to river bed lowering and subsequent salt water ingression. Later, a new pumping station was commissioned in 2000, 16 km inland from the river mouth. Recently, this pumping station also had to be shifted further upstream (i.e., 18 km from the river mouth point) because of water quality problems. All these indicate the high environmental cost incurred as a result of channel incision induced aggravated salt water intrusion in the Nilwala river due to indiscriminate sand mining and subsequent channel incision. As a result of the ever increasing environmental effects of sand mining, people's opposition has heightened in midlands and lowlands of the major rivers of Sri Lanka. Sri Lanka has a long history of people mediated campaigns showcasing collective actions in cases related to environmental governance.

In Sri Lanka, sand is also mined from river mouths and dunes for construction purposes. In some rivers such as the Kelani river, sand is being mined at rates far greater than the natural replenishment. The low sand dunes along the coast of Uswetakeiyawa area have also been heavily mined leading to instability in the entire area (Lowry and Wikramaratne 1988). Beach sand mining is concentrated mainly in the coastal segment between Talpitiya and Panadura on the west coast of the island. Beach sand is used for road construction, land reclamation, and also for filling foundations of buildings.

The construction industry is the main consumer of river sand and accounts for more than 95 % of the demand. It is also estimated that almost one third of the total sand demand in Sri Lanka is met from illegal sources. It is now well understood that the river sources of sand alone is inadequate to meet the increasing national demand. Offshore sand, land sand, dune sand and manufactured sand are some of the alternate sources of construction sand in Sri Lanka whose estimated availability are 31.5 million m^3, 9.6 million m^3, 0.3 million m^3, and 9.96 million m^3, respectively. The increased price of offshore sand makes it less popular compared to river sand. The presence of shells and chlorides in marine sand is another disadvantage, hintering its wide use in building constructions.

5.9 Nepal

The growing trend of urbanization in Nepal has enhanced the demand of sand for building constructions. Now the demand has been met by mining the aggregate materials from terraces and river beds. Illegal sand mining is reported from many parts of the country (Sayami and Tamrakar 2007). Although river bed excavation is prohibited, illicit extraction of sand is continuing in the rivers of Central Nepal, especially in the Sanla and the Manahara rivers. It is estimated that about 3,123 m^3 of sand is being extracted daily (1,885 m^3 from river and 1,238 m^3 from terraces) for various purposes. Studies reveal that there is an imminent need to strengthen laws and policies in the country to make mining activities environment-friendly. As the quality of sand is poor because of the presence of mica (10–32 %) and other flacky minerals, proper processing is required to improve sand quality within acceptable limits before its end use in the construction industry. The volume of sand extracted from instream and terrace sources of different river basins of Nepal is given in Fig. 5.4.

5.10 Maldives

In Maldives, an island nation in the Arabian Sea, sand extraction for building constructions is from the lagoons at depths of up to 3 m below sea water level. About 6 tonnes of calcareous sand is collected by a single boat in one trip. Specific locations are chosen for extraction of sand with different grain sizes. Sand is the most important building material in the island nation, in terms of quantity. Annual sand production is about 62,000 m^3 (Brown 1997). About 60 % of the sand is used for building constructions and the remaining for other purposes.

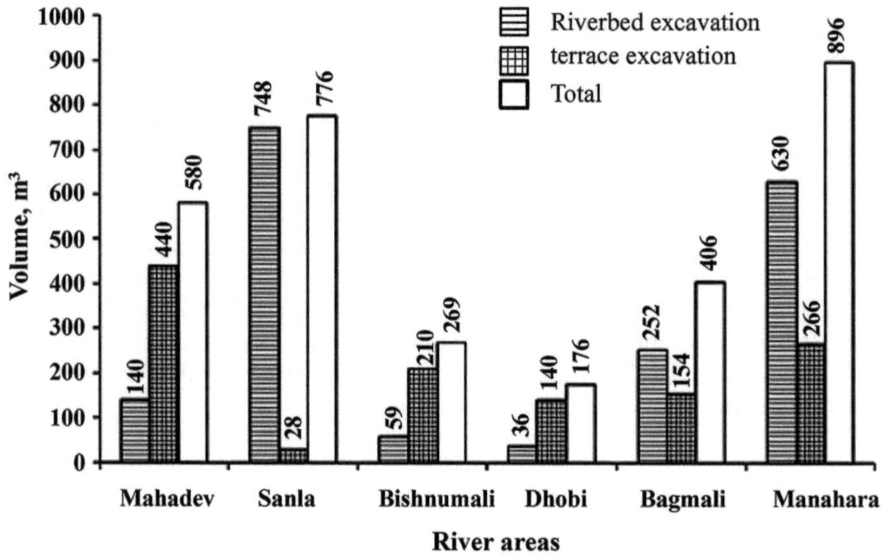

Fig. 5.4 Volume of sand extracted from river and terrace sources of some of the important rivers of Nepal (Sayami and Tamrakar 2007)

5.11 India

Huge volumes of sand and gravel are extracted from the river systems of India every year for meeting the ever increasing demand of aggregate materials for building constructions. As urbanization and settlements expand, the demand for aggregates (especially river sand and gravel) have also increased. As seen from Fig. 5.5, it is very evident that the major source of revenue collection the mineral industry in Kerala state in the south west coast of India was from the minor mineral sector, especially from sand from alluvial sources. But, only a few documentations exist in literature that deals with the problems of river sand mining. One such study was made by Mohan (2000) in Kulsi river, Assam, which categorically reveals that river sand mining is one of the potential threats to the dolphin population in the river. In another study, Sridhar (2004) reported that sand mining in Coleroon river (tributary of Cauvery river), Tamil Nadu causes serious environmental problems to the river system. In a study, Ronnie (2006) reported about the widespread illegal sand mining on the banks of Shimsha River near Kokkare, Bellur in Bangalore (Karnataka State). The study reveals that the sand mining activity has already affected the life of avian fauna of that region because of various reasons including the clearing/loss of riparian vegetation due to unabated sand mining. Recently, Hanamgond (2007) made an attempt to document river sand mining from different parts of the country—Chapora creek (Goa), Kali estuary (Karwar, Karnataka) and Karli estuary (Konkan coast, Maharashtra). In an article, Saviour (2012) has made a review of the environmental impacts of soil and

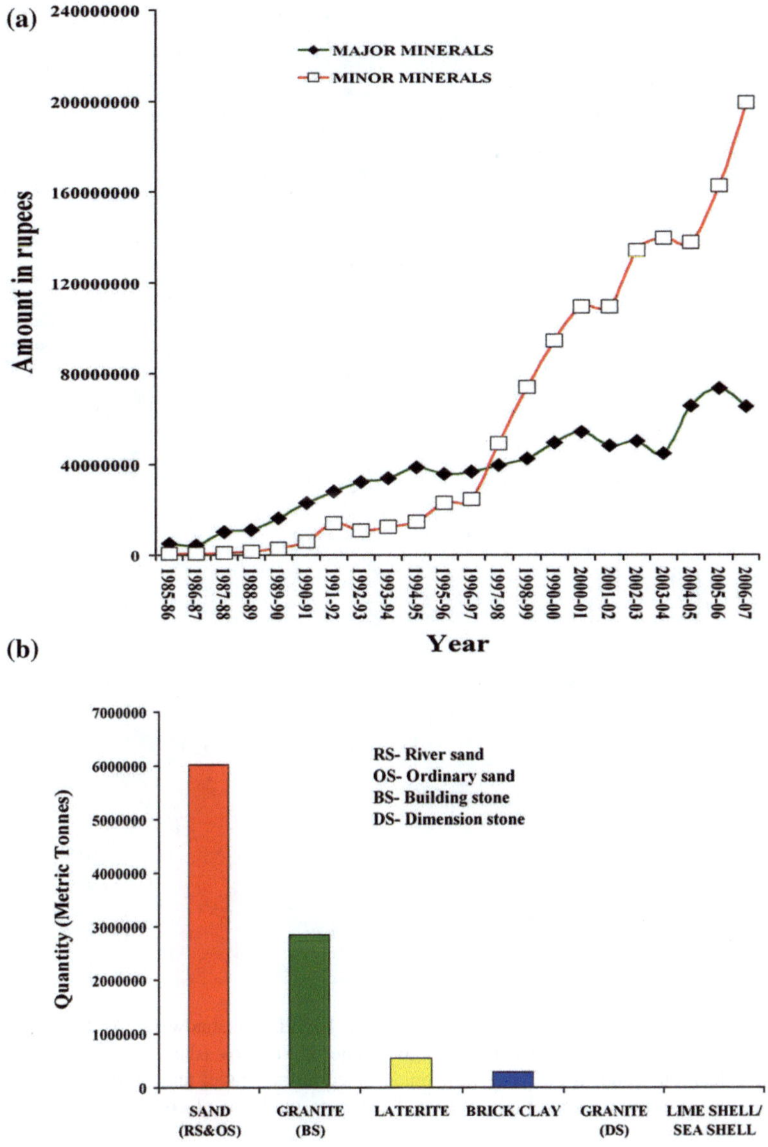

Fig. 5.5 Revenue collection from the extraction of minerals (major and minor) in Kerala. **a** Revenue collection for the period 1985/86–2006/07; **b** production of minor minerals during 2006/07. (*Data source* Department of Mining and Geology, Government of Kerala)

sand mining from different parts of the country. Studies of Sonak et al. (2006) made a review of sand mining activity, rules and regulations and the resulting environmental issues of the rivers of Goa. In Goa, sand mining is being carried out in a large scale. Most of the operations are illegal and has incurred huge loss of

Plate 5.1 Calcareous sand and gravel of Minicoy Island, Lakshadweep Archipelago. **a** An extensively developed sandy beach in the lagoon-side of Minicoy island, **b** a view of coralline gravel, **c** calcareous sand, **d** building wall made of coralline gravel and mortar, **e** a compound wall made up of coralline gravel and mortar, **f** concrete made up of coralline gravel, sand and cement

riverine sand resources and environmental degradation of the rivers in the state. In Tamil Nadu, floodplain sand mining permitted in private lands adjacent to river beds has enabled private owners to encroach on the river bed illegally. Public roads were also seriously damaged. Direct irrigation to about 22,000 acres of land was affected in Vaigai and Cauvery basins (Sonak et al. 2006).

In Laccadives, coral sand and gravel extracted from beaches, lagoons, and inland areas are used traditionally for building constructions (Plate 5.1). However,

Plate 5.2 Evidences of river bed changes. Scenarios of Kuzhikkala *kadavu* of Manimala river during **a** 1999 and **b** 2009. Note the two *circles* in the plate. In the first case, the people could cross the river because of the shallow bed level. In the second case **b** the people are unable to cross because of channel incision. Channel incision at this section is over 1.5 m; **c** The person standing marks (approx.) the channel bed level in 1989. The local self government has to construct eight more steps in 2009 to reach the new bed level. The maximum channel incision could be $\sim$1.6 m

in recent years, building materials brought from the nearby areas of the mainland (i.e., Kerala) is also used for building constructions.

The environmental problems of river sand mining are severe in the small rivers i.e., the rivers with catchment area <10,000 km^2 (Milliman and Syvitzki 1992) draining the western flanks of the Western Ghats. Although the river bed resource is very limited in their storage zones, the quantity of sand mining is manifold higher than natural replenishments. In Kerala, the environmental effects of river sand mining have been a theme of research from 1986 onwards with the pioneering investigations of Thrivikramaji (1986). He addressed the environmental consequences of sand mining from one of the important southern most rivers of Kerala, the Neyyar river. Later, Padmalal (1995), Padmalal et al. (1999) carried out a series of studies in various rivers of Kerala with specific reference to river sand mining.

The environmental impacts of sand and clay mining in the Vamanapuram river basin were carried out by Saritha et al. (1996). A study on the sand budget of Periyar river with special reference to river sand mining was made by Padmalal and Arun (1998). CESS (1998) conducted a study on sand mining in Manimala river falling within the jurisdiction of Mallapally grama panchayat, Pathanamthitta district and pointed out that the river has markedly deteriorated due to indiscriminate scooping of sand from its active channels. The studies of CESS and Centre for Water Resources Development and Management (CWRDM) undertaken as per the direction of Hon'ble courts and Government of Kerala were instrumental in formulating the legislation "The Kerala River banks Protection and Regulation of Removal of sand Act, 2001" (Kerala Gazete 2002). Apart from the physical aspects of sand mining, a few studies are carried out on the biological aspects of sand mining as well. An attempt to study the impact of sand and gravel mining on the benthic fauna of the Achankovil river flowing through the Pathanamthitta district was made by Sunilkumar (2002b). The sand mining activities in the Manimala river and its adjoining floodplains were documented by Arun et al. (2003). In a study on the Ithikkara river (Kollam district), Sheeba and Arun (2003) opined that, out of the 25 freshwater fishes recorded in the river, a total of 16 species were under threat, mainly, due to habitat loss resulted from sand mining (Table 5.3). Chandramoni and Anirudhan (2004) documented the impacts of land sand mining in the downstream parts of the Muvattupuzha river. The impact of sand mining on the physical and biological environment of river systems of Kerala was dealt by Arun et al. (2006). Padmalal et al. (2008) carried out an analysis on the environmental effects of river sand mining from the hinterland drainages of Vembanad lake and summarized the major environmental effects of river sand mining (Table 5.4). As determination of the rate of channel incision is a difficult task in areas devoid of river gauging stations. Padmalal (2010) used "environmental proxies" for understanding channel incision (Plate 5.2). Sreebha and Padmalal (2011) made an environmental impact assessment of river sand mining in the rivers draining into the Vembanad lake and suggested a set of management plans to enhance the overall environmental quality of the rivers. Studies conducted by Prasad (2011) in Bharathapuzha river revealed that clandestine instream and

Table 5.3 Threatened fresh water fishes of Ithikkara river and reasons for threat as per IUCN (1990); *Source* Sheeba and Arun (2003)

Sl. no.	Name of fish	Threat	IUCN category
1	*Anabas testudineus*	Damming, fishing, human interference, over exploitation, trade (local, domestic, commercial)	VU
2	*Barilius bakeri*	Fishing, loss of habitat, pesticides, poisoning, siltation, trade (local)	VU
3	*Channa striatus*	Fishing, tradc (commercial)	LRlc
4	*Clarias batrachus*	Trade (commercial, local, domestic)	VU
5	*Danio aequipinnatus*	Human interference, loss of habitat, pollution, trade (local, domestic)	LRnt
6	*Hyporhamphus xanthopterus*	Fishing, human interference, loss of habitat, over exploitation, pesticides, poisoning, trade (local, domestic)	CR
7	*Hypselobarbus curmuca*	Disease, dynamiting and other destructive method of fishing, loss of habitat due to exotic animals, over exploitation, predation, predation by exotic animals, trade (local, domestic)	EN
8	*Lepidocephalus irrorata*	Human interference, Fragmentation, Trade (local)	VU
9	*Noemacheilus pulchellus*	Damming, fishing, human interference, loss of habitat, over exploitation, poisoning, pollution, trade (local)	DD
10	*Parluciosoma daniconius*	Fishing, pollution, trade (local, domestic)	LRnt
11.	*Puntius sarana sarana*	Fishing, human interference, loss of habitat, trade (local, domestic)	
12	*Puntius ticto punctatus*	Fishing, genetic problems, over exploitation, trade (local)	CR
13	*Puntius vittatus*	Fishing, human interference, loss of habitat, over exploitation, pollution, trade (commercial)	VU
14	*Wallago attu*	Decline in prey species, hunting for food, poisoning, siltation, trade (local, domestic, commercial)	LRnt
15	*Xenentodon cancila*	Fishing, pollution, trade (domestic)	LRnt
16	*Glossogobius giuris*	Hunting, trade (local)	LRnt

VU Vulnerable
LRlc Lower Risk least concern
LRnt Lower Risk near threatened
CR Critically Endangered
EN Endangered
DD Data Deficient

flood plain sand mining is detrimental to the infiltration wells and galleries of the rural water supply schemes as the removal of sand bed/inter layers within the river channel and its flood plains could adversely affect the net groundwater storage and summer flows (lean flows) through the river. Further, reduced flows through the river channel can disturb the fresh water—saline water balance in the areas close

Table 5.4 The general impacts of river sand and gravel mining on various components of river ecosystem

Sl. no.	System/ Components	Impact of mining
1	River channel	Erosion of riverbanks; riverbank slumping; lowering of river channels; changes in riverbed configuration; undermining of engineering structures like bridges, water intake structures, side protection walls, spillways, etc.; loss of placer mineral resources associated with alluvial sand and gravel
2	Surface water	Rise in suspended particulate level, turbidity and other pollutants like oils, grease, etc., from vehicles used for the removal of sands; ponding of water and reduction in natural cleansing capacity of river water; aggravated salt water ingression
3	Groundwater	Lowering of groundwater table in areas adjacent to mining sites; damage to the fresh water aquifer system in areas close to the river mouth zones
4	Flora and fauna	Dwindling of floral and faunal diversity within river basin; decline in terrestrial insects like mayfly, dragon fly, stone fly etc., whose larval stages are in the shallow water sandy fluvial systems; habitat damage/ loss and changes in breeding and spawning grounds; reduction in inland fishery resource
5	Culture	Damage to culturally significant places; places of annual religious congregations, etc
6	Coast/ nearshore	Lack of replenishment of coastal beaches leading to coastal erosion and reduction in the supply of nutrient elements from terrestrial source

to the river confluence zones. This will aggravate the salt water intrusion problems in the coastal lowlands. Studies of Maya et al. (2011) in the Muvattupuzha river basin also reiterated the adverse impacts of widespread flood plain sand mining (Fig. 5.6) in the geo-hydrological setting of the area. It is estimated that an amount of 2.689 million tones per year of flood plain sand is quarried from the lowlands of the Muvattupuzha river basin from 76 active pit mines (Table 5.5) mainly using high power jet pumps (Plate 5.3). The process of pumping and indiscriminate sand mining aggravates the salt water intrusion problems in the area, turning many fresh water dug wells saline.

Beach sand has been quarried in certain places of Gulf of Mannar, Andaman islands, Lakshadweep islands, etc. Surveys of sub littoral deposits of sand by the Geological Survey of India around the most southerly Minicoy atoll in the Laccadives revealed deposits of calcareous sand up to 288 million tonnes at a depth of 1 m in the lagoon. Recently, the Marine and Coastal Survey Division of the Geological Survey of India has identified five promising areas in the offshore waters of Kerala with an estimated resource of 2,030 million tonnes of marine sand resource suitable for use in construction industry (Dinesh et al. 2014).

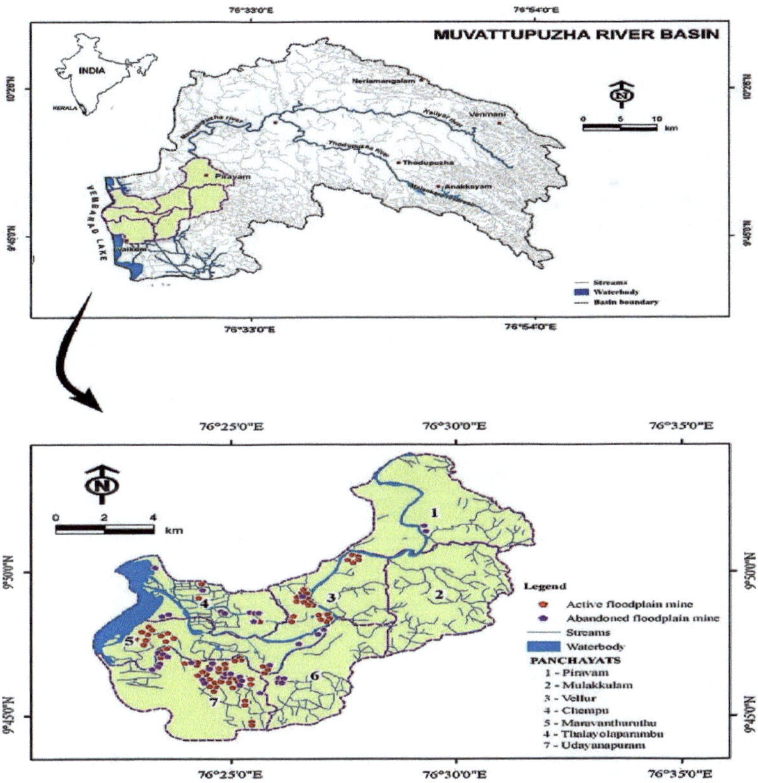

Fig. 5.6 Floodplain mining locations in the Muvattupuzha river basin

Table 5.5 Details of floodplain mining in the Muvattupuzha river basin

Sl. no.	Local body	Number of floodplain mines		Quantity mined from active sites ($\times 10^6$ ty−1)	Number of labourers
		Active	Abandoned		
1	Piravom	–	2	–	–
2	Vellur	24	1	0.780	984
3	Thalayolaparambu	–	4	–	–
4	Chembu	3	11	0.162	285
5	Maravanthuruthu	16	4	0.612	670
6	Udayanapuram	33	19	1.135	1383
Total		**76**	**41**	**2.689**	**3322**

5.12 Summary

Lack of availability of published papers on river sand mining is a major challenge for addressing this very important environmental issue that threatens the very existence of river ecosystems. In most countries, indiscriminate river sand and

Plate 5.3 Mechanised (using high power jet pump) and manual floodplain mining from the coastal lowlands of Muvattupuzha river basin, Southwest India

gravel mining imposes irreparable damages to the river ecosystems including its instream and floodplain areas. The main environmental adversities include channel incision, valley widening, loss of riparian vegetation, head cutting, and undermining of bridges. Lack of suitable, low cost alternatives to river sand resource is a problem that favour over extraction of river sources of sand and gravel. Studies carried out by eminent experts in this area show that small rivers are most affected due to indiscriminate sand mining as the river bed resources in such rivers are limited. The developed countries are now inclined towards rock based fine and coarse aggregate materials for meeting the demand of construction sector. Rocks like limestone, different varieties of igneous and metamorphic rocks, and sandstone are used for the production of aggregate materials. As the processes are energy-intensive and incur high cost of production, many countries still depend on the river sources of sand for construction purposes and hence are unable to ban river sand mining although the activity is a threat to river ecosystems.

References

Alvarado-Villalon F, Harrison DJ, Steadman EJ (2003) Alluvial aggregate mining in Costa Rica. British Geological Survey Commissioned Report, CR/03/50N, p 27

Arun PR, Mini SR, Reghunath R, Padmalal D (2003) Mining of construction grade fine aggregates from river basins: a case analysis. In: Proceedings of the 15th Kerala science congress, Thiruvananthapuram, pp 586–590

Arun PR, Sreeja R, Sreebha S, Maya K, Padmalal D (2006) River sand mining and its impact on physical and biological environments of Kerala rivers, Southwest Coast of India. Eco-chronicle 1:1–6

Ashraf MA, Maah MJ, Yusoff I, Wajid A, Mohmood K (2011) Sand mining effects, causes and concerns: a case study from Bestariyaya, Selengor, Peninsular Malaysia. Sci Res Essays 6(6):1216–1231

Athukorala K, Navaratne C (2009) Sand barriers and access to water community pressure and policy interventions in river sand mining in Sri Lanka. In: Proceedings of IWMI, Battaramulla, Sri Lanka

Baker BH (1954) Geology of the Southern Machakos district. Report No. 27, Geological Survey, Kenya

Brown BE (1997) Integrated coastal management: South Asia. Department for International Development (DFID), United Kingdom

Bull WB, Scott (1974) Impact of mining gravel from Urban Stream Beds in the Southwestern United States. J Geol pp 171–174

California Department of Water Resources (1994) Use of alternate gravel sources for fishery restoration and riparian habitat enhancement, Shasta and Tehama counties, Northern District, Red Bluff, CA, p 189

California Division of Mines and Geology (1987) Mines and mineral producers active in California during 1986. California Division of Mines and Geology. Spl Publ 93:57

California Division of Mines and Geology (1994) Surface Mining and Reclamation Act of 1975

Castoglioni GB, Pellegrini GB (1981a) Two maps on the dynamics of a river bed. Erosion and sediment transport measurement. In: Proceedings of the IAHS symposium, Florence, 22–26 June 1981, pp 223–228

Castoglioni GB, Pellegrini GB (1981b) Geomorphologia dell'alveo del Brenta nella pianura tra Bassano e padova. In: Zunica M (ed) II territorio della Brenta. Ammo. Prov. Di Padova-Universita di Padova, Padova, pp 12–32

Castro J, Cluer B (2003) Instream aggregate mining issues in Oregon. White paper prepared for the US fish and wildlife service, Oregon fish and wildlife office, Portland, OR, p 22

CESS (1998) Sand availability of the Manimala river in the vicinity of Mallapally grama panchayat, Pathanamthitta district: a technical study. Centre for Earth Science Studies, Thiruvananthapuram, p 9

Chandramoni CM, Anirudhan S (2004) Land sand mining: a new environmental challenge. In: Proceedings of the Indian environment congress, Thiruvananthapuram, pp 142–147

Diang'a (1992) Dangers of wanton sand harvesting. Daily Nation, 24 April, p 11
DID (2009) River sand mining and management guideline. Department of Irrigation and Drainage, Malaysia, p 50
Dinesh AC, Praveen Kumar P, Shareef NM and Jayaprakash C (2014) Marine sand resources of Kerala coast vis-à-vis acute shortage of construction sand in the State of Kerala. In: Shaji E and Pradeep Kumar AP (eds) Mineral resources of Kerala. Proceedings of Mineralia 2014, University of Kerala, Trivandrum, India, pp 93–105
ECD (2000) Environmental impact assessment (EIA) guidelines for river sand and stone mining. State Environmental Conservation Department, Sabah, Malaysia
Erskine WD, Geary PM, Outhet DN (1985) Potential impacts of sand and gravel extraction on the Hunter river, New South Wales. Aus Geo. Stu 23:71–86
Farrant AR, Mathers SJ, Harrison DJ (2003) River mining: sand and gravel resources of the lower Rio Minho and Yallas fan-delta. Jam Brit Geol Surv, Commissioned Report, CR/03/101N, p 24
Hanamgond PT (2007) Sand mining photographs. J Coast Res 23:949–950
IUCN (1990) Red list of threatened animals. International union for the conservation of nature and natural resources. World Conservation Monitoring Centre, Cambridge
Kerala Gazette (2002) The Kerala Protection of River Banks and Regulation of Removal of Sand Act (2001). Kerala Gazette, vol 47, no 285, Government of Kerala, p 14
Kitetu J, Rowan J (1997) Integrated environmental assessment applied to river sand harvesting in Kenya. In: Patric CK, Lee N (eds) Sustainable development in a developing world—integrated socio-economic appraisal and environmental assessment, Edward Elgar, Cheltenham, pp 189–199
Kondolf GM, Swanson ML (1993) Channel adjustments to reservoir construction and in-stream gravel mining, Stony Creek, California. Environ Geol Water Sci 21:256–269
Kondolf GM (1997) Hungry water: effects of dams and gravel mining on river channels, Environ Manag 21: 533–551
Lamberti A (1993) Le modificazioni recenti verificatesi nellasta principale del Po e problemi connessi. Acqua Aria 6:589–592
Leeuw JD, Shankman D, Wu G, Boer MFD, Burnham J, He Q, Yesou H, Xiao J (2010) Rategic assessment of the magnitude and impacts of sand mining in Poyang Lake, China. Reg Environ Change 10:95–102
Lu XX, Zhang SR, Xie SP, Ma PK (2007) Rapid channel incision of the lower Pearl River (China) since the 1990s as a consequence of sediment depletion. Hydrol Earth Syst Sci 11:1897–1906
Luo XL, Zeng EY, Ji RY, Wang CP (2007) Effects of in-channel sand excavation on the hydrology of the Pearl River Delta, China. J Hydrol 343:230–239
Maya K, Padmalal D, Santhosh V, Aneesh SR (2011) Instream and flood plain sand mining from Muvattupuzha river basin, consequences and strategies. In: Proceedings of the national seminar on mining of river sand and its impacts on the environment, 18–19 Feb 2011, Centre for Water Resources and Management, Kozhikode, India, pp 111–124
Meador MR, Layher AO (1998) Instream sand and gravel mining: environmental issues and regulatory process in the United States. Fisheries 11:6–13
Milliman JD, Syvitski JPM (1992) Geomorphic/tectonic control of sediment discharge to the ocean: the importance of small mountainous river. J Geol 100:525–544
Mohan RSL (2000) The 'Blind' river dolphins of India. http://www.grieb.org/dolphins
Padmalal D (2010) Alluvial sand mining: the Kerala experience. Commissioned papers—Western Ghats. http://westernghatsindia.org/commssioned-papers
Padmalal D (1995) A preliminary investigation on the sand budget of the Pamba river between Cherukolpuzha and Kizhavarakadavu. Centre for Earth Science Studies, Thiruvananthapuram, p 6
Padmalal D, Arun PR (1998) Sand budget of Periyar river: special reference to river sand mining. Centre for Earth Science Studies, Thiruvananthapuram, p 55

Padmalal D, Maya K, Arun PR (1999) Sand mining from the Pamba river: consequences and strategies. Centre for Earth Science Studies, Thiruvananthapuram, p 30

Padmalal D, Maya K, Sreebha S, Sreeja R (2008) Environmental effects of river sand mining: a case from the river catchments of Vembanad lake, Southwest Coast of India. Environ Geol 24:879–889

Padmalal D, Maya K, Baijulal B (2011) Environmental appraisal and sand auditing of Periyar river, Ernakulam district, Kerala. Centre for Earth Science Studies, Thiruvananthapuram, Kerala, pp 1–212

Prasad NBN (2011) Impact of sand mining on water supply scheme – a case study. In: Proceedings of the national seminar on mining of river sand and its impacts on the environment, 18–19 February 2011, Centre for Water Resources and Management, Kozhikode, India, pp 153–159

Priyantha AAC, Mayooran D (2011) On the mineral resources in Vavuniya district. Report No. MR/G/65/2010, Geological Survey of Mines Bureau, Dehiwda, p 52

Ranjana UKP (2011) River sand mining and associated environmental problems in Sri Lanka. In: Walling D (ed) Sediment problems and sediment management in Asian River Basins, International Association of Hydrological Science, United Kingdom, pp 148–153

Ranjana UKP, Champa MN, Athukorela K (2009) River sand mining and environmental issues in Sri Lanka. Joint international convention of 8th IAHS scientific assembly and 37th IAH congress, Hyderabad, India

Rinaldi M (2003) Recent channel adjustments in alluvial rivers of Tuscany, Central Italy. Earth Surf Proc Land 28:587–608

Rinaldi M, Simoncini C, Sogni D (2005) Variazioni morfologiche recenti di due alvei ghiaiosi appenninici: il F. Trebbia ed. il F. Vara. Geografia Fisica e Dinamica Quaternaria Suppl vol 7, pp 313–319

Ronnie (2006) Illegal sand mining affecting bird life. http://bengaluru.yulop.com

Saritha D, Sajikumar S, Machado T (1996) Sand and clay mining in the Vamanapuram river basin and its environmental impacts. In: Proceedings of the 8th Kerala science congress, Kochi, p 20

Saviour MN (2012) Environmental impact of soil and sand mining: a review. Int J Sci Environ Technol 1:125–134

Sayami M, Tamrakar NK (2007) Status of sand mining and quality in Northern Kathmandu, Central Nepal. Bulletin of the Department of Geology, Tribhuvan University, Nepal, vol 10, pp 89–98

Sheeba S, Arun PR (2003) Impact of sand mining on the biological environment of Ithikkara River: an overview. In: Proceedings of 15th Kerala science congress, Thiruvananthapuram, pp 806–807

Sonak S, Pangam P, Sonak M, Mayekar D (2006) Impact of sand mining in local ecology. In: Sangeetha S (ed) Multiple dimensions of global environmental change. The Energy Research Institute, New Delhi, p 324

Sreebha S, Padmalal D (2011) Environmental impact assessment of sand mining from the small catchment river in the Southwestern Coast of India. Environ Manage 47:130–140

Sridhar L (2004) Cauvery delta: New reality. http://www.indiatogether.org/2004/mar/env-groundh20.htm

Sunilkumar R (2002b) Impact of sand mining on benthic fauna: a case study from Achankovil River: an overview. Catholicate College, Pathanamthitta district, Kerala, p 38

Surian N, Rinaldi M (2003) Morphological response to river engineering and management in alluvial channels in Italy. Geomorphology 50:307–326

Surian N, Rinaldi M, Pelligrini L, Audisio C, Maraga F, Teruggi LB, Turitto O, Ziliani L (2009) Channel adjustments in northern and central Italy over the last 200 years. Geol Soc Am Spec Pap (in press)

Suzanne RF (2002) Instream gravel mining and related issues in Southern Missouri. USGS Fact Sheet 012-02

Thrivikramaji KP (1986) River metamorphosis due to human intervention in Neyyar river basin. Final Technical Report, Department of Environment, Government of India, p 112

Wang Z, Ding J, Yang G (2012) Risk analysis of slope instability of levees under river sand mining conditions. Water Sci Eng 5:340–349

Wikramaratne HJM (1988) Coastal area management in Sri Lanka. In: Coastal management. The University of Chicago, Chicago, pp 263–293

Chapter 6
Environmental Case Studies from SW India

Abstract Kerala state in the southwestern coast of India is blessed with 41 west flowing rivers that debouch into the Arabian Sea. These rivers are small with limited river bed resources. Among the various human interventions that are threatening rivers, indiscriminate river sand mining is of paramount importance as the activity irreparably degrades rivers and its adjoining aquatic environments. Studies reveal that the river bed in their storage zones is lowering at a rate of 5–20 cm per year. In the present chapter, two case studies have been examined to disclose the severity of environmental problems of sand mining from the rivers draining the Western Ghats—an ecologically sensitive area in the Peninsular India.

Keywords Kerala rivers · Manimala river · Periyar river · Sand mining · Environmental problems

6.1 Introduction

Blessed with heavy monsoon rains and a sloping topography of Western Ghats, the southwestern coast of India, especially falls in the Kerala state, is drained by 41 west flowing rivers. These rivers are small with river length varying between 16 and 244 km and the catchment area being <6,200 km^2. The Kerala rivers originate from different head water elevations (70–2,066 m). Out of the total rivers, 19 are mountainous (head water elevation 1,000–3,000 m above msl), six are upland (500–1,000 m above msl), and eight each are lowland (100–500 m above msl) and coastal plains (<100 m above msl) rivers, as per the classification of Milliman and Syvitski (1992). The major mountainous and upland rivers are depicted in Fig. 6.1. Many of these rivers originate from the Western Ghat mountains and flow through the highland (>75 m amsl), midland (75–8 m amsl), and lowland (<8 m amsl) physiographic provinces of the state (PWD 1974; CESS 1984). Although river bed resources are very limited, the rivers in Kerala are subjected to indiscriminate extraction of sand and gravel from their instream and floodplain areas for meeting the demand in the ever increasing construction sector.

D. Padmalal and K. Maya, *Sand Mining*, Environmental Science and Engineering,
DOI: 10.1007/978-94-017-9144-1_6,

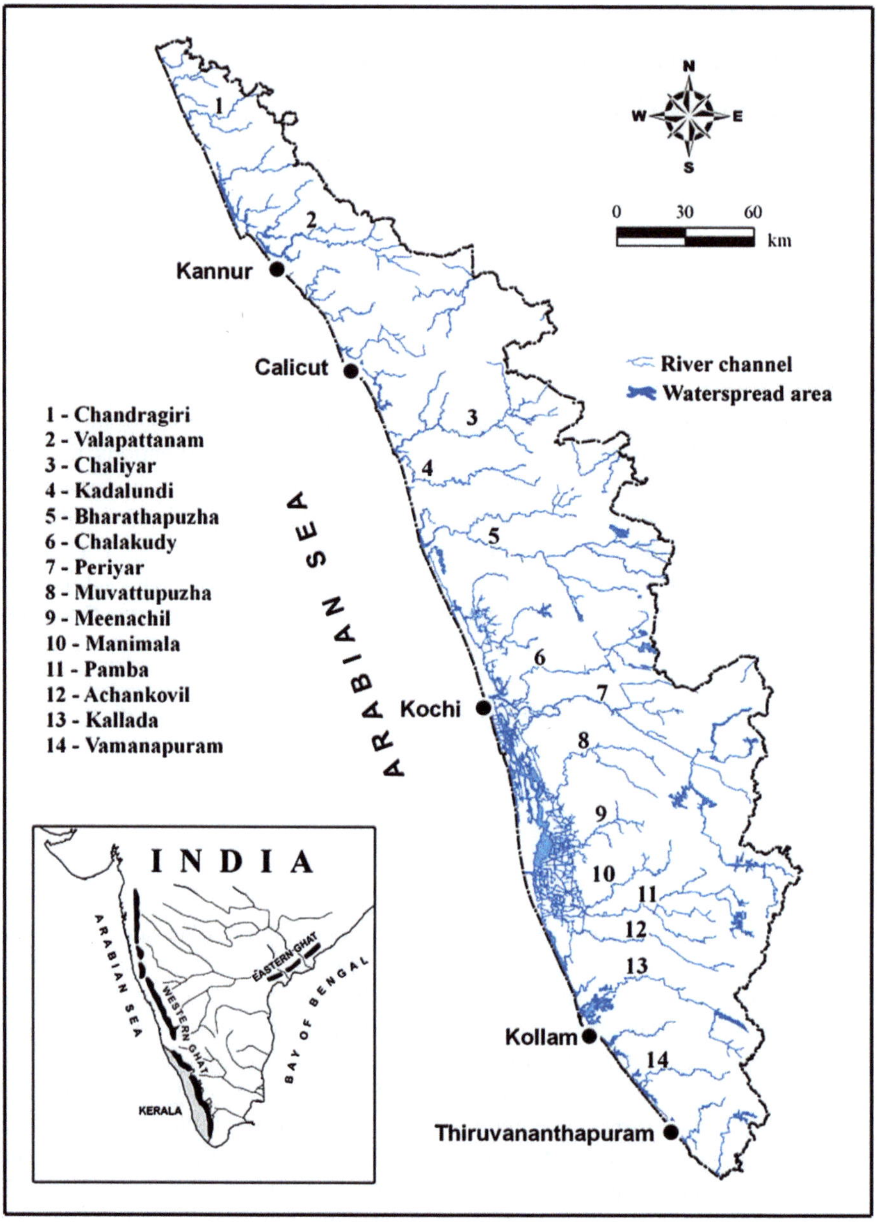

Fig. 6.1 Important rivers of Kerala in the southwest coast of India. *Note* Inset map of Peninsular India showing the Western Ghats mountains from which the small west flowing rivers are originating

River sand mining gained momentum in the state since the first half of the 1970s consequent to the boom in construction sector fuelled by the rise in foreign remittance and per capita income generation. Although the intensity of mining was

Table 6.1 Mountainous and upland rivers of Kerala along with their length, basin area, and sand mining details (see Fig. 6.1 for locations of the rivers)

Sl. No.	River name	Length (km)	Basin area (km²)	Annual sand extraction ($\times 10^6$ m³)		Annual sand replenishment[a] ($\times 10^6$ m³)	Nos. of SMS[b]	Bed lowering[a] (cm/year)
				Total	Below CWC-GS[a]			
1	Chandragiri	105	1,406	0.515	0.240	0.045	35	NA
2	Valapattanam	110	1,867	0.648	0.200	0.065	43	5
3	Chaliyar	169	2,923	1.086	0.164	0.077	73	10
4	Kadalundi	130	1,122	0.422	0.409	0.009	90	16
5	Bharathapuzha	209	6,186	1.426	0.271	0.026	153	5
6	Chalakudy	130	1,704	0.460	0.282	0.006	45	13
7	Periyar	244	5,398	3.470	2.005	0.041	319	18
8	Muvattupuzha	121	1,554	0.980	0.537	0.013	260	7
9	Meenachil	78	1,272	0.140	0.043	0.004	72	15
10	Manimala	90	847	0.660	0.100	0.009	153	12
11	Pamba	176	2,235	0.420	0.095	0.014	64	15
12	Achankovil	128	1,484	0.500	0.240	0.006	68	9
13	Kallada	121	1,699	0.320	0.199	0.046	108	11
14	Vamanapuram	88	687	0.280	0.111	0.004	65	11

[a] Recorded at the Central Water commission's gauging station (CWC-GS)
[b] *SMS* sand mining sites
NA not available; Achankovil is an upland river; all others are mountainous

low in the 1970s and 1980s, the activity became a potential threat to the river ecosystems in the second half of the 1990s due to the ripple effects of the developmental processes to which the region has been subjected. In Kerala, sand mining is reported from almost all the rivers and nearby aquatic environments. Table 6.1 shows the quantity of sand mining and other relevant details of some of the important (mountainous and upland rivers) rivers of Kerala. The locations of these rivers are depicted in Fig. 6.1. Indiscriminate mining has lowered the river bed of Pamba upto 3–4 m during the past three decades. In certain stretches, the river bed has turned into deep pools with depth reaches up to 6 m or more. This has serious implications on the fresh water availability and biological diversity of the river basin environment. It is estimated that an amount of 30 million tons of river sand has been extracted from the Kerala rivers during the period 2005–2006 from more than 2000 sand mining locations (Fig. 6.2). Despite numerous prohibitions and regulations, sand mining continues to be a major environmental threat in Kerala state. Water table in the riparian areas/riparian wells on either side of the river channel has lowered markedly due to indiscriminate sand mining. And, as a result, the land once known for its plentiful rice harvest in the floodplain areas faces severe scarcity of water for raising paddy. The paddy cultivation has now declined alarmingly as the lowlands are exploited for sand and brick/tile earths.

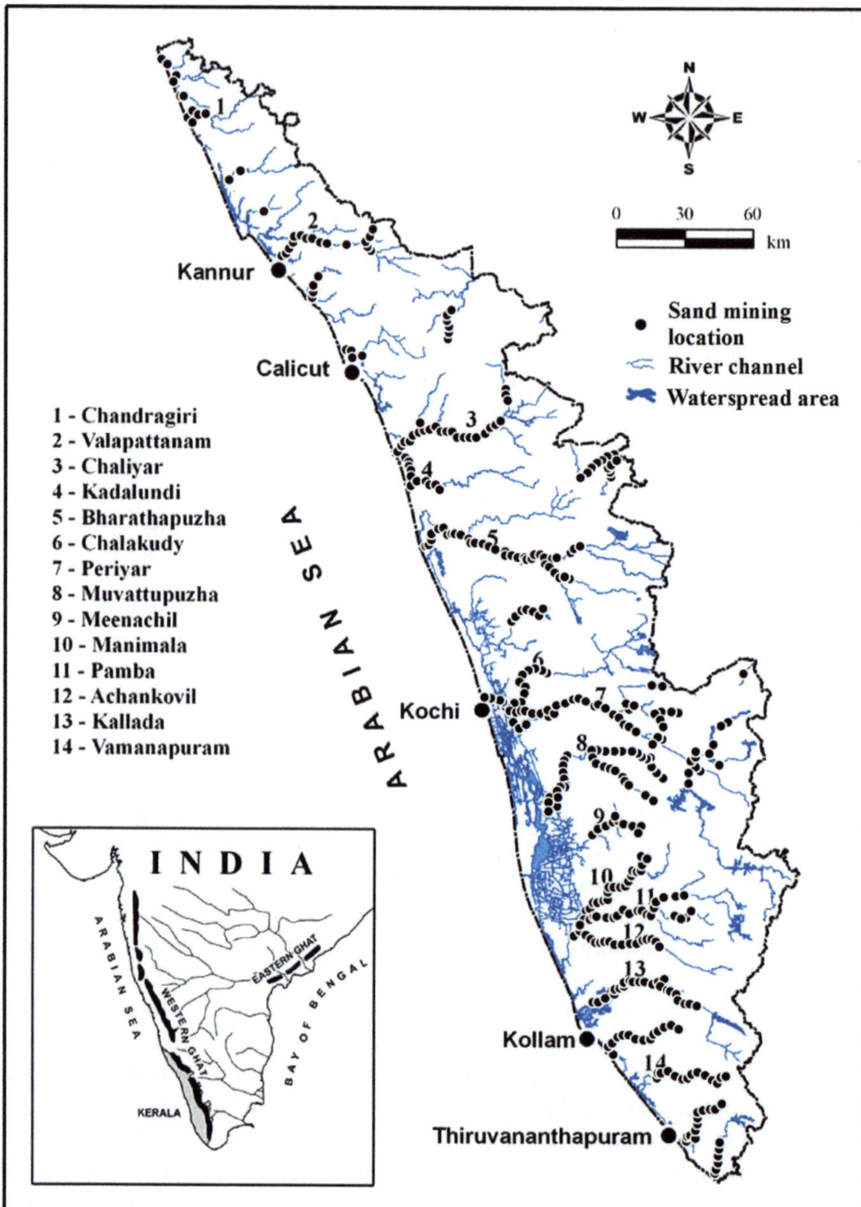

Fig. 6.2 Rivers of Kerala showing sand mining locations

Sand and gravel are classified as a "minor" mineral in terms of its economic value, despite being dredged in huge proportions from the river systems. This has grave environmental impacts on the mining-hit rivers of the state that have already recorded river bank instability, damages to engineering structures and river ecology (Padmalal et al. 2008), as well as drastic reduction in agricultural productivity (Shiekha et al. 2014). This is one side of the problem. On the other side, lack of adequate availability of construction grade sand or any other alternatives creates constraints on in the developmental sector.

One way of overcoming the resource deficit in construction sector is to import sand from regions having its surplus availability. As many of the environmental effects of river sand mining are similar to that of other human interventions, like construction of dams and embankments, it is often difficult to discriminate the environmental effects of river sand mining from such activities. In this chapter, the cases of two important rivers of Kerala—Manimala and Periyar rivers—are examined to address the environmental issues culminated from sand mining and their causative factors in the small rivers of Kerala in particular and the developing economies in tropical regions in general. Based on the areal extent of the river basins, Rao (1973) classified Indian rivers into three broad categories: (1) minor rivers (river basin area <2,000 km^2), (2) medium rivers (2,000–20,000 km^2), and (3) large rivers (>20,000 km^2). According to this classification, Manimala river falls under the minor river category, whereas Periyar river in the medium river category.

6.2 Manimala River

The Manimala river (Fig. 6.3) is a small river (minor river as per, Rao 1973) in southern Kerala which was perennial before the 1970s. But now a greater part of the river in its upstream has turned to be ephemeral during summer season. Indiscriminate sand mining and subsequent lowering of the river bed and landuse changes have had a significant stake in the observed change. The river has a length of about 90 km and catchment area of 847 km^2 (CWRDM 1995). The Manimala river originates from the Western Ghats at an elevation of 1,156 m above MSL and merges with the river Pampa near Thiruvalla before finally emptying into the Vembanad lagoon (the largest backwater system in the west coast of India). The river basin falls under all three physiographic divisions of Kerala—lowland, midland, and highland. The major tributaries of the river are Kokkayar, Pullaga *Ar*, and Kanjirappalli *thodu*. Manimala river is devoid of any reservoirs/dams in its catchments, but the river is dissected at many places by check dams. The river exhibits a dendritic to sub-dendritic drainage pattern. The total population and population density of the basin are 7.78 lakhs and 736 inh. km^{-2}, respectively (Census of India 2001). The river basin area comprises a spectrum of landuse classes which includes plantations (41.48 %), agricultural lands (39.02 %), built-up area (9.63 %), forests (8.32 %), barren land and shrub lands (4.23 %), water bodies (0.98 %), and grasslands (0.06 %) (Kerala State Landuse Board, Government of Kerala). The mean monthly discharge through the river was monitored at three stations, each located at upstream (Mundakkayam station), midstream

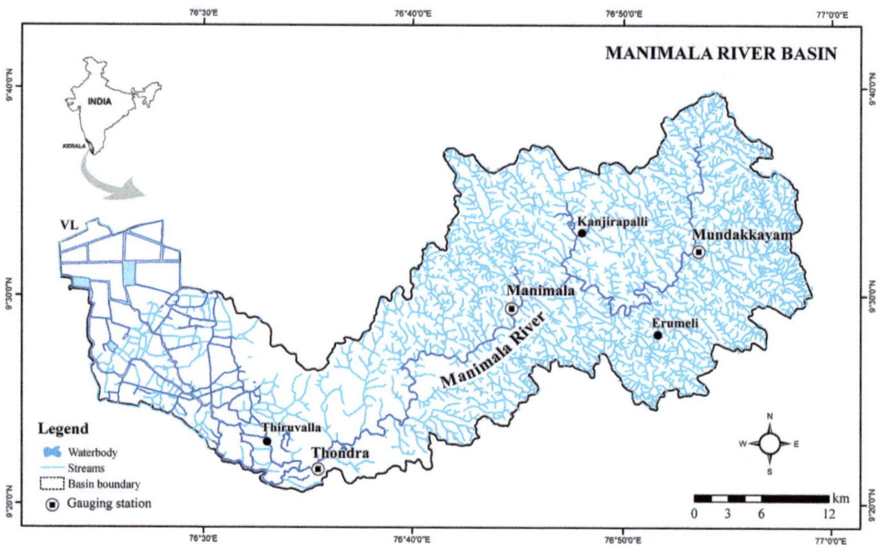

Fig. 6.3 Manimala river showing drainage characteristics. *VL* Vembanad lagoon

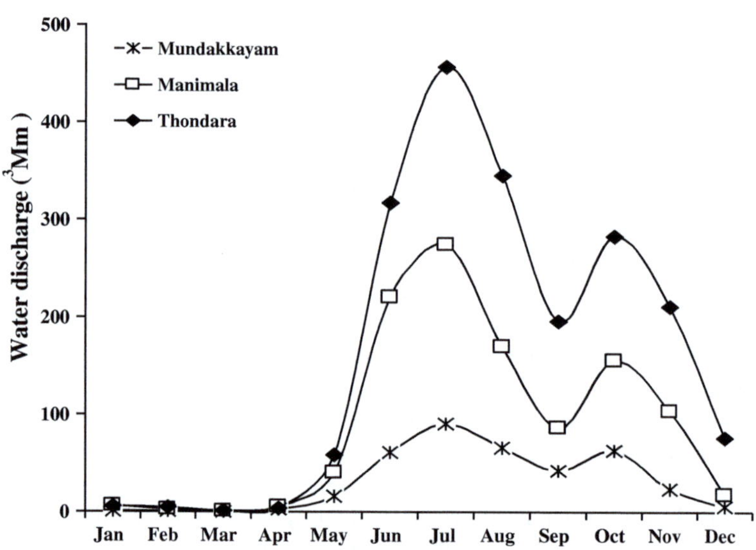

Fig. 6.4 Mean monthly discharge of water through Manimala river during the period 1990–2000. The peak discharge in July is due to southwest monsoon and October is due to northeast monsoon (see Fig. 6.3 for location of the gauging stations)

(Manimala station), and downstream (Thottumukham station) areas (Fig. 6.4). The river discharge is influenced by southwest (June–September) and northeast (October–December) monsoons.

6.2.1 Instream Sand Mining

As rampant sand mining activities in the river impose serious environmental problems in the basin environments, the Centre for Earth Science Studies (CESS), a premier research institution, based at Thiruvananthapuram in Kerala state was directed by the Hon'ble Courts and the Government of Kerala to carry out studies on the impacts of sand mining on various environmental components of Kerala rivers. One of the rivers where such a study has been taken up was Manimala river. Although sand mining was widespread in the river in the past 2–3 decades, not much information was available for scientific computations. CESS estimates reveal that a total of 2.47×10^6 ty^{-1} of sand was extracted from the river channel during 2002–2003 period. In the Manimala river, sand and gravel mining was particularly intense during 2004–2005 period. The entire master channel of the river and its major tributaries and distributaries were exploited for vast volume of sand and gravel. The physiography-wise split up on the quantity of sand mining from Manimala during 2004–2005 is given in Table 6.2. Sediment extraction (instream and floodplain) was widespread in the river stretch falling within midland (0.913×10^6 ty^{-1}) compared to the other two physiographic zones (highland 0.079×10^6 ty^{-1}; lowland 0.07×10^6 ty^{-1}). Eighteen local bodies located on either sides of the Manimala river were engaged in instream sand mining. Apart from the master channel, sand mining was reported from the small tributaries (0.042×10^6 ty^{-1}) as well. Five local bodies in highlands were engaged in river sand mining. Spatial analysis reveals that about 74 % of the extracted sand is from the midland reaches. The sand extraction from the highlands and lowlands were about 13 and 12 %, respectively. A total of 166 sand mining locations were active along the river, with a higher number from midland (127) compared to highland (32) and lowland (7) (Fig. 6.5).

6.2.2 Floodplain Sand Mining

Apart from instream mining, a substantial quantity of sand is being mined from the floodplain areas of the Manimala river as well. Increased demand of construction grade sand, depletion or near absence of sand in many stretches of the active channel, stringent rules and regulations, periodic ban of instream mining, increased awareness regarding the negative effects of instream sand mining, etc., are some of the factors that have promoted floodplain sand mining in the river basin. It is estimated that about 91,200 tons of sand is being scooped out annually from the floodplain areas of the Manimala river basin (Arun et al. 2006). It is well understood that the sand-rich layers interbedded within floodplain deposits have a great bearing on the local hydrology and ground water movement in the area (Fig. 6.6). Water logging in pits left after mining creates many adverse effects such as collapse of structures adjacent to mine sites and marked changes in the landuse pattern and ultimate decline in the net agricultural productivity of the region.

Table 6.2 Quantity of sand mining (QSM), labour force (LF), and number of sand mining sites (SMS) in different physiographic zones of Manimala river

District/physiography	Number of local bodies	QSM ($\times 10^6$ ty^{-1})		LF (Number)		SMS (Number)	
		T/D	MC	T	MC	T	MC
Highland	5	0.039	0.04	187	197	22	10
Midland	11	0.003	0.91	10	1,642	1	126
Lowland	2	–	0.07	–	232	–	7

T tributary; *D* distributary; *MC* main channel; *ty*$^{-1}$ tons per year

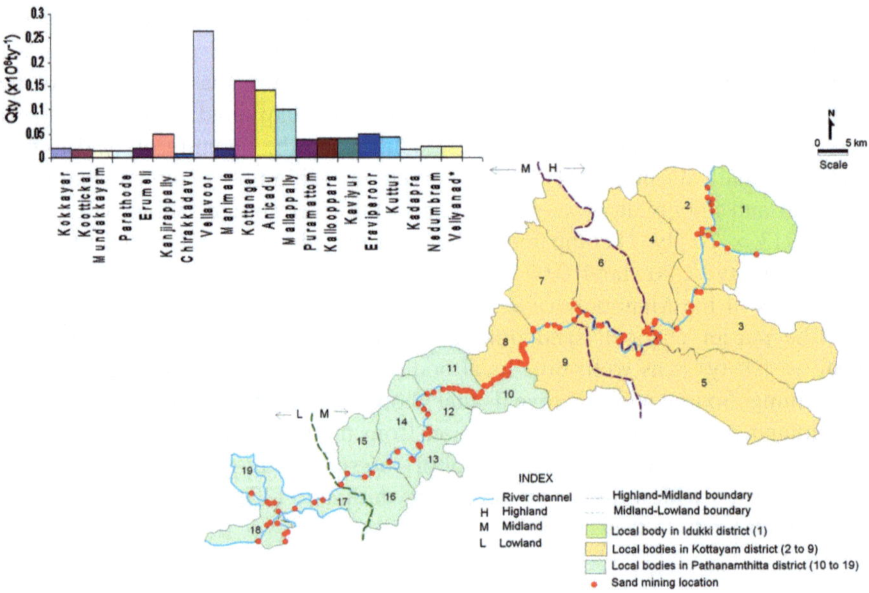

Fig. 6.5 Sand mining locations and local bodies engaged in sand mining from the Manimala river basin

6.2.3 Channel Incision

The records of river cross-profile measurement of Central Water Commission gauging stations provide ample evidences of river bed changes/channel incision during the past 2–3 decades. The shape of the river cross-section has narrowed and deepened steadily for the last few decades as the channel was superimposed onto older alluvium. The cross-profile measurements of the Manimala river revealed that the channel bed at many locations in the storage zone has lowered by several meters due to indiscriminate sand mining over the years. At a gauging station located 20 km upstream of the river confluence (with Pamba river), the river bed has lowered to 169 cm during the last 18 years with an annual average lowering of

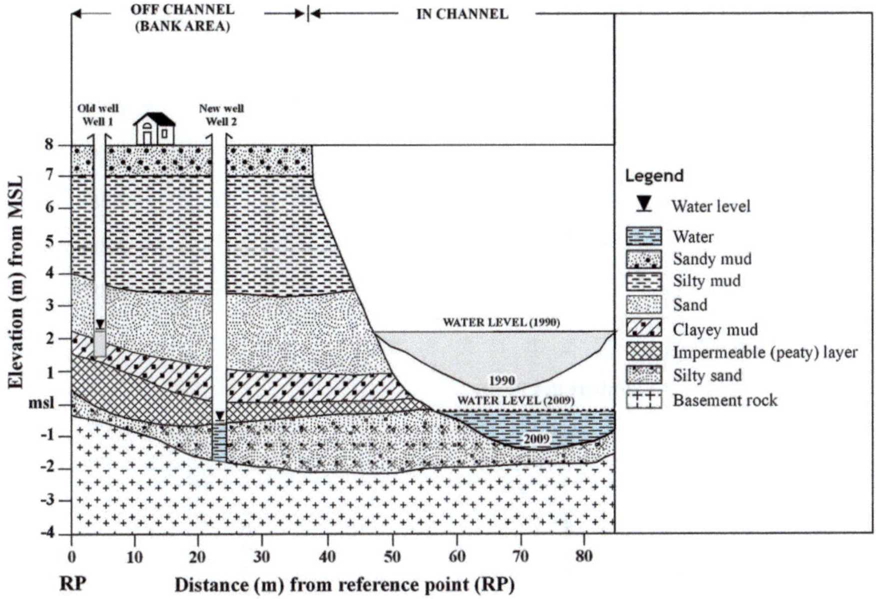

Fig. 6.6 Depletion of well water is a direct consequence of sand mining and river bed lowering. A schematic model of an area near Karippumuri *kadavu* showing changes in water level positions in relation to channel incision

9.4 cm y^{-1} (Fig. 6.7). The rate of the lowering of this section was 13.84 cm y^{-1} during the period 1990–1995. But during the next 5-year period (1995–2000), the channel incision reduced substantially to the tune of 8.14 cm y^{-1}. During 2000–2008, the channel incision again increased to 11.82 cm y^{-1} due to rampant mining for sand and gravel. River channel in the storage zones at many places have turned into deep pools/pits (Fig. 6.8). The river bed lowering and subsequent fall in sand deposits adversely affected the ground water table in the areas adjoining the banks. Plate 6.1 shows a few evidences of channel incision in Manimala river. A total of 11 bars and point bars in the storage zone of the Manimala river disappeared due to indiscriminate sand mining in the past four decades. River widening due to channel incision is also severe at many places in the alluvial reach (Plate 6.1).

6.3 Periyar River

Periyar river is the longest river in Kerala in the southwestern coast of India (Fig. 6.9). The river has a length of 244 km and a catchment area of 5,398 km^2 (CWRDM 1995). The river originates from the Western Ghat mountains at an elevation of about 1,830 m amsl and flows through highly varied geologic and geomorphic terrains. The major tributaries of the river are Muthirapuzha *Ar*,

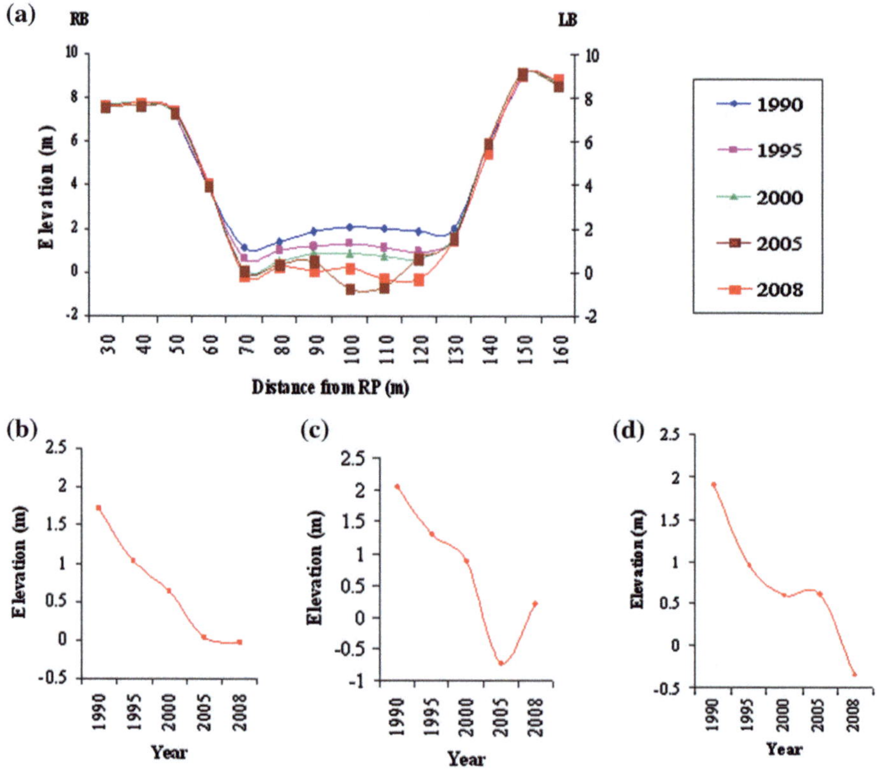

Fig. 6.7 Bed lowering of Manimala river during 1990–2008. **a** Average bed lowering. **b** Bed lowering at 100 m from RP. **c** Bed lowering at 120 m from RP; *LB* left bank; *RB* right bank; *RP* reference point

Perinjankutty *Ar*, Idamala *Ar*, and Mangalapuzha *Ar*. The river has two distributaries—the Mangalapuzha distributary and the Marthanda Varma distributary. The population and population density in the highland, midland, and lowland physiographic units are 857,266 and 194 inh. km^{-2}, 560,923 and 805 inh. km^{-2}, and 20,567 and 9,050 inh. km^{-2}, respectively. The minimum rate of flow of the river is 9.66 m^3/sec and the maximum is 1,364.66 m^3/sec (Joy 1992); (Fig. 6.10). The flow of the river is highly restricted by dams constructed in its highland reaches. There are 16 hydroelectric projects and nine irrigation schemes attached to the river Periyar, which regulate discharge to significant levels.

6.3.1 Sand Mining from Periyar River

Periyar river is the lifeline of Central Kerala. The river plays an important role in the economic, religious, traditional, and cultural heritage of the state. But the river

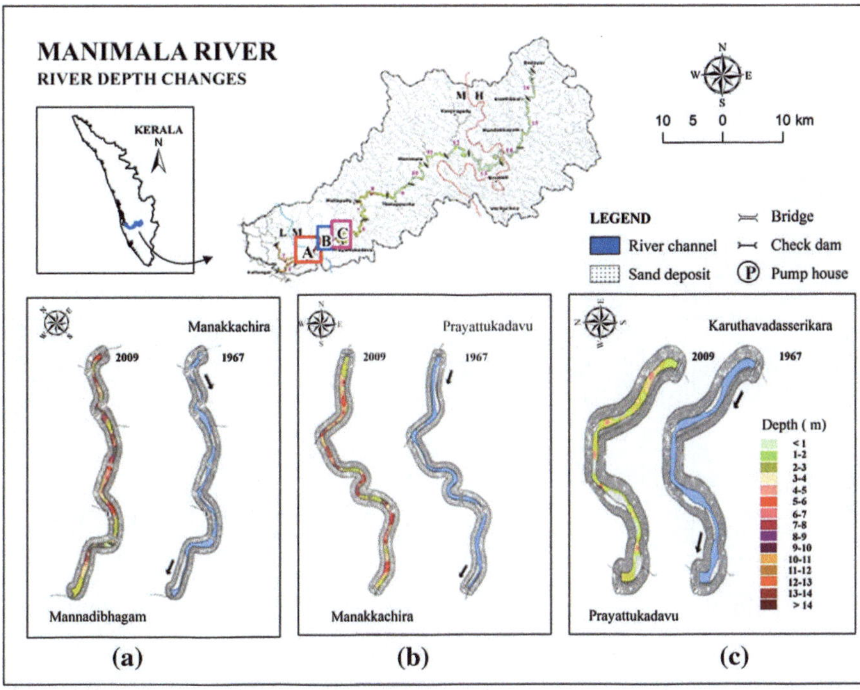

Fig. 6.8 River depth changes in certain stretches of Manimala river, SW India

is now on the verge of severe deterioration due to indiscriminate scooping of sand over the years. Periyar river is subjected to rampant sand mining all along its course including the main channel as well as the tributary/distributary systems (Plate 6.2). The annual instream sediment extraction from Periyar was 7.636×10^6 ty^{-1}, in which 5.657×10^6 ty^{-1} was extracted from the main channel and 0.979×10^6 ty^{-1} from tributary–distributary systems. A major share of sand extracted is from the midland part of the river (5.315×10^6 ty^{-1}) and the rest is from the highland (0.707×10^6 ty^{-1}) and lowland regions (0.653×10^6 ty^{-1}). Various local bodies of Idukki, Ernakulam, and Thrissur districts of the state located on either side of the river were engaged in sand mining from the river, with the highest demand met by Ernakulam district, which is a fast developing region in the state. The locations of sand mining in Periyar river are given in Fig. 6.11. A greater proportion of sand is extracted from the midland part of the Periyar river, in spite of less number of sand mining locations, compared to the other two physiographic zones—the highlands and midlands. This is in direct response to the rising demand of sand for the fast developing urban centers in the midland area (Fig. 6.11). The detailed account of sand mining from the three physiographic zones of Periyar river is given in Table 6.3.

Plate 6.1 Impact of sand mining and channel incision on the riparian environment. **a** A house is under threat due to unabated sand mining. The >5 m wide courtyard of the house has eroded by hungry water during high flow regime of southwest monsoon; **b** channel widening due to sand mining is a common scene in the lowlands of Manimala river; **c** and **d** collapse of side protection structures due to river bed lowering; **e** collapse of duricrusts resulted from undermining of sand and subsequent channel incision; **f** river bank slumping and uprooting of riparian vegetation due to channel incision

Twelve local bodies were engaged in sand mining from the midland part of the river. The maximum quantity of sand extracted by a local body in the midland was 0.96×10^6 ty^{-1}, whereas the minimum was estimated to be 0.019×10^6 ty^{-1}. An amount of 0.653×10^6 ty^{-1} of sand was extracted from the two distributary channels of the river, namely Mangalapuzha (0.533×10^6 ty^{-1}) and Marthandavarma (0.12×10^6 ty^{-1}) distributaries. Sand mining activity was noticed in the

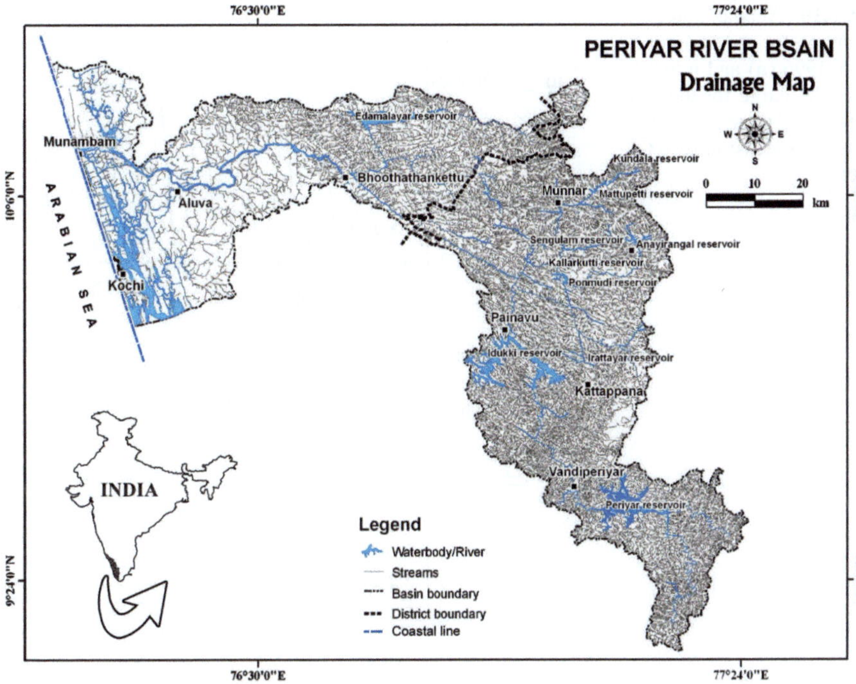

Fig. 6.9 Periyar river showing drainage characteristics

highland local bodies of the Periyar river basin, though in much lesser quantities compared to the midland regions. Out of 19 local bodies involved in sand mining in the highland region, 12 local bodies extract sand from the alluvial deposits of the Munnar–Peermedu plateau. A total amount of 0.177×10^6 ty^{-1} of sand was extracted from the tributaries and 0.15×10^6 ty^{-1} from the main river channel in the highland areas. Local bodies in the highlands other than plateau region extract about 0.147×10^6 ty^{-1} of sand from the tributaries and 0.192×10^6 ty^{-1} from the main channel. Among the 12 local bodies engaged in sand mining from the plateau region, seven local bodies extract sand from the tributaries (0.157×10^6 ty^{-1}) and three from the main channel (0.104×10^6 ty^{-1}). Among the two physiographic zones, the midlands are the most affected, as the quantity of sand mined is about seven to eight fold higher than that in the highlands and lowlands.

In addition to mining of sand and gravel from the active channels (instream mining), a significant quantity of sand has been extracted from the floodplains/overbank areas of the river as well. Floodplains and terraces (former floodplains) are the sites of sediment storage and contain large quantities of sand and gravel within it. The floodplain sand is extracted usually by pit excavation method. Pits in the floodplains often deepen below the water table as it provides a convenient water source for the separation of desirable size of aggregate materials. In Periyar

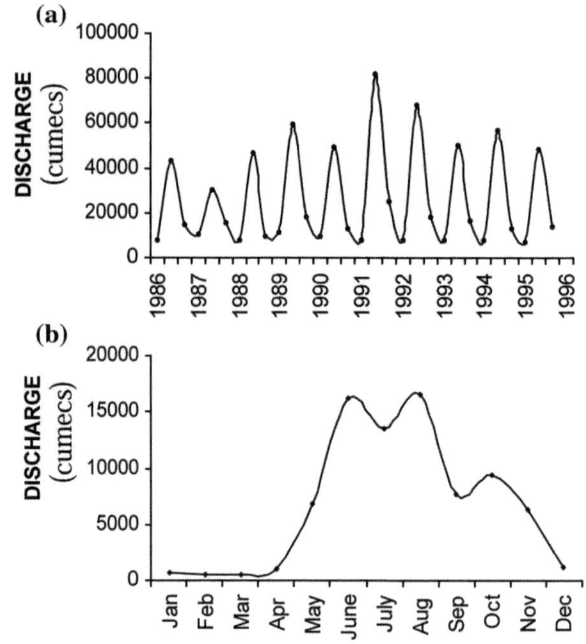

Fig. 6.10 Annual water discharge for the period 1986–1996 (**a**) and monthly water discharge for the year 2004 (**b**); *Source* Central Water Commission (CWC), Kochi

river basin, floodplain mining is noticed in a few locations in the lowlands. Additionally, land sand mining from the paddy lands is noticed in the plateau region of the highlands. The Rapid Reserve Estimation (RRE) survey conducted by CESS during the period 1998–1999 revealed that the storage zone of Periyar accounted for a maximum sand reserve of about 53.4×10^6 tons.

6.3.2 Channel Incision

The Central Water Commission's (CWC) gauging station at Neeleswaram located about 50 km upstream of the confluence of Periyar river with the Arabian Sea, records continuously the changes in water quality and quantity since early 1980s. The deepest point in the channel cross-sectional profile was sited at 85 m from the reference point of CWC and river bed at this station has lowered to 740 cm, with an estimated average channel incision of 24.66 cm y^{-1}, during 1980–2010. A high-resolution analysis of the cross-profile measurements shows that the rate of channel incision was 4.1 cm y^{-1} during 1980–1985 and 3.64 cm y^{-1} during 1985–1990. In the subsequent 5-year period, the rate of incision has increased exponentially turning the river channel into deep pools. The increase in the observed channel incision was about three times the higher during 1990–1995 (13.2 cm y^{-1}) and seven times the higher during 1995–2000 (28.64 cm y^{-1}) with respect to the incision noticed during 1985–1990 period. During the period

Plate 6.2 Some selected scenes from the sand mining sector. **a–c** Sand mining from Periyar river; **d** pole and scooper used for the extraction of sand from deep channel areas; **e** loading of sand in carrier vehicles, and **f** an array of boats ready for sand extraction

2000–2005, the channel incision rate had reduced considerably to 9.64 cm y^{-1}, with respect to the previous 5-year period (i.e., 1995–2000) (Fig. 6.12). However, the average channel incision during 2005–2010 was quite high (19.28 cm y^{-1}). The deepest point also showed the same trend of the average channel incision. Figure 6.13 depicts the sand bed changes with respect to mean sea level at Neeleswaram gauging station and with respect to bed rock at Malayattoor.

The Malayattoor station is located about 4.5 km upstream of the Neeleswaram CWC gauging station. A sand bar measuring a maximum thickness of about 4 m in 1998 vanished almost completely, reaching the bed rock level in the year 2005.

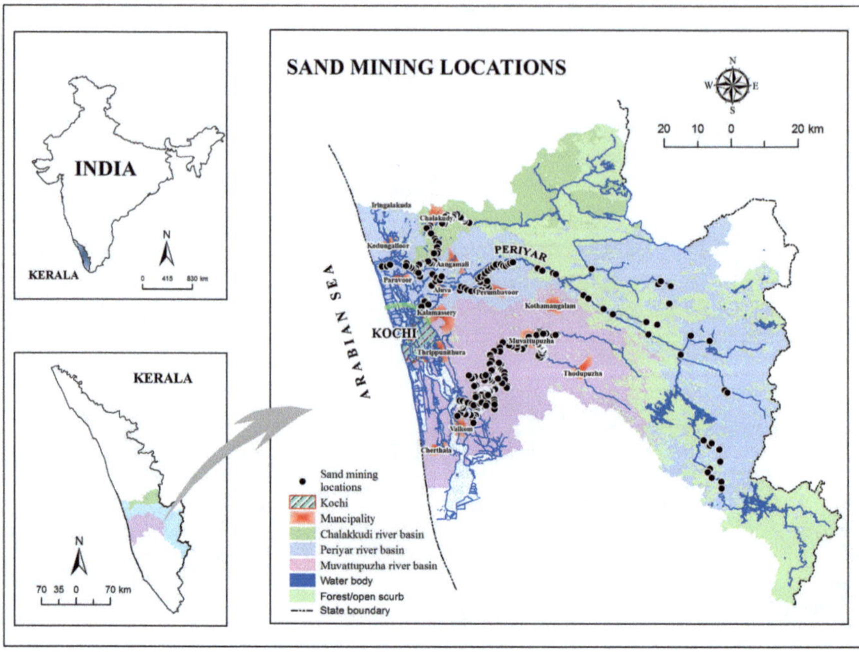

Fig. 6.11 Sand mining locations in Periyar river along with that of Chalakudy and Muvattupuzha river draining through the central part of Kerala state, SW India. A substantial part of the sand extracted from the downstream reaches is used for meeting its demand in the construction sector of Kochi city and its satellite townships

Table 6.3 Quantity of sand mining (QSM), labour force (LF), and number of sand mining sites (SMS) in the Periyar river

District/physiography	Number of local bodies	QSM ($\times 10^6$ ty^{-1})		LF (Number)		SMS (Number)	
		T/D	MC	T	MC	T	MC
Highland	19	0.326	0.33	899	906	129	63
Midland	12	–	5.31	–	750	–	100
Lowland	9	0.653	–	2,060	–	65	–

T tributary; *D* distributary; *MC* main channel; *ty*$^{-1}$ tons per year

6.4 Drivers of Change

Urban development is one of the major drivers of environmental change across the earth surface. In the third quarter of the twentieth century alone, urban population increased over 100 % worldwide and nearly 200 % in less developed regions (Gupta 1984). As might be expected, the development of infrastructure to accommodate expanding populations poses formidable demands on river resources

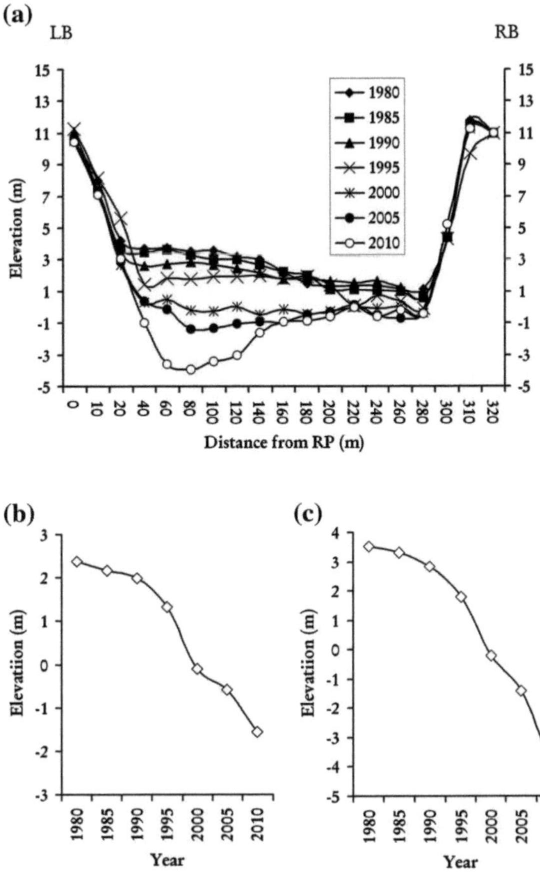

Fig. 6.12 Channel cross-profile changes of Periyar river recorded at the Neeleswaram CWC gauging station. **a** River bed lowering of Periyar river during 1980–2010; **b** average bed lowering estimated for the entire section; and **c** bed lowering noticed at the deepest point of the channel at 80 m away from RP of CWC gauging station. *LB* left bank; *RB* right bank; *RP* reference point; elevation is measured with respect to mean sea level (MSL) (*Data source* Central Water Commission (Kochi))

(Eyles 1997; Douglas 2005), a situation that is likely to continue into the future as well. According to Rubin (2001), the three prominent factors determining environmental changes are population rise, standard of living, and technological progress. Standard of living of the population is usually measured in terms of percapita gross domestic product. The more affluent the population (indicated increase in per capita income), the more goods and services required to satisfy their needs and hence greater will be the resulting environmental impacts and degradation (Fig. 6.14). Third critical factor is technology—the vehicle for delivering goods and services that people demand. These three factors are closely interrelated with one another and are principal drivers that determine the future of landuse requirements, availability of natural resources, emission of pollutants to air, water and land, etc. (Fig. 6.15), (Rubin 2001; Sreebha 2008). The environmental impacts of human activities, including sand and gravel mining, depend directly on the number of people inhabiting an area.

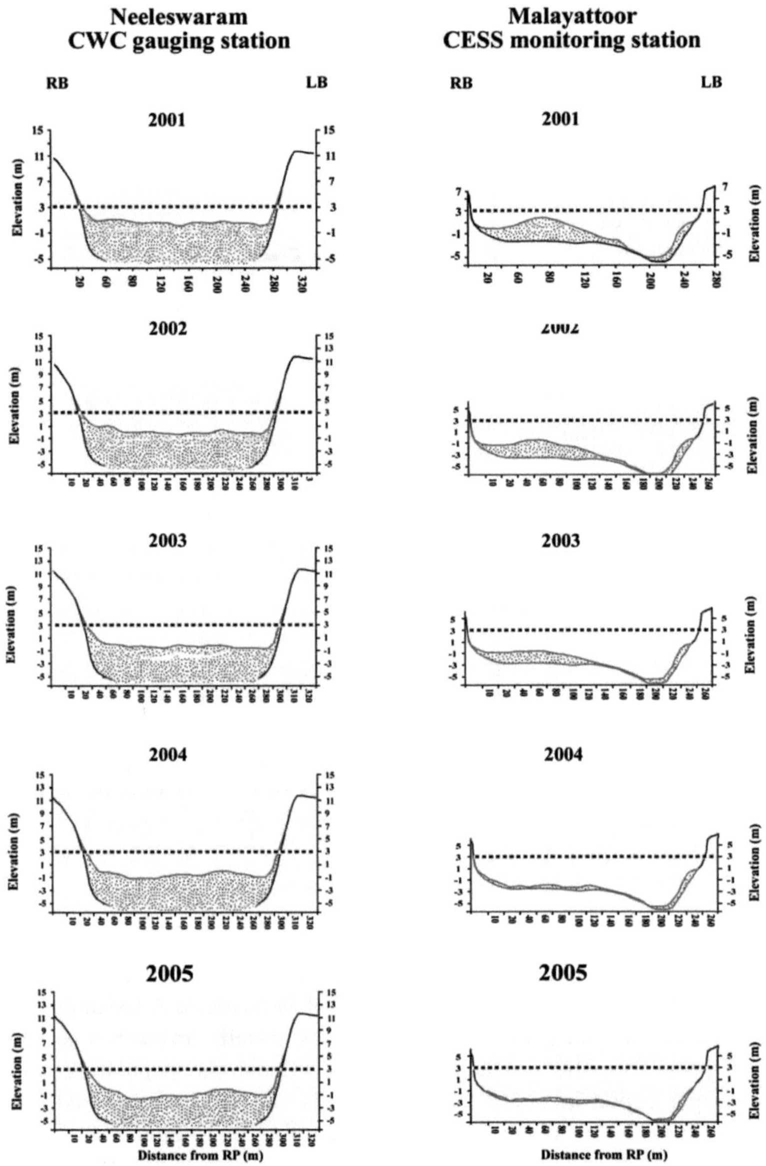

Fig. 6.13 Channel cross-profile changes noticed in the Periyar river measured at the gauging stations of CWC during 1985–2004. Elevation is measured with respect to mean sea level (*Data source* Central Water Commission, Government of India)

The interplay of the three main drivers of environmental change gains considerable significance in Kerala state located in southwestern India. Figure 6.16a shows the decadal addition of houses and population growth in the state since the

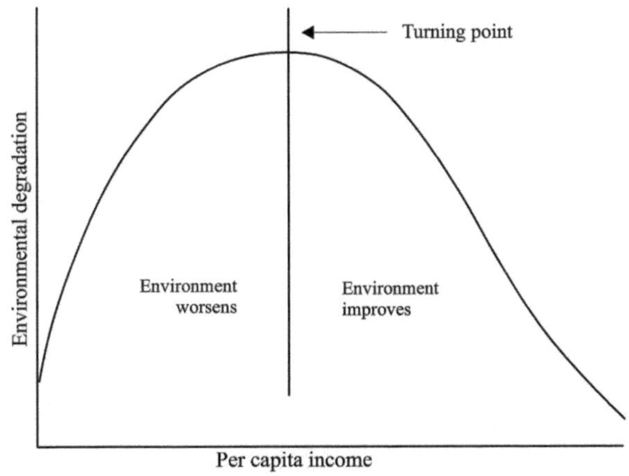

Fig. 6.14 Kuznets curve—an inverted "U". Although the wealth of the environment worsens during the first half of the development process, it will improve in the second half because of the strict rules and regulations as well as change in perception of the people towards sustainable development

beginning of the last century. The rise in foreign remittance was one of the most influential factors responsible for the increase in the residential sprawl in Kerala state (Fig. 6.16b). Indiscriminate resource extraction for construction of houses and other infrastructural facilities has already made notable disturbances in the state's natural environmental setting which is reflected well in the aquatic systems of the region, especially the rivers.

However, estimates show that the growth in population will attain stability within a couple of decades (Kannan 2004) as the population control measures in the state are yielding significant positive responses. As shown from Table 6.4, the decadal addition of persons in a square kilometer area is found to decrease regularly from 1971 onwards. This is an indication of the efficiency of the population control system in the state. The population will be reaching a stable figure of around 35 million (currently 31 million) by around early 1930s (Kannan 2004). It is quite interesting to note that Kerala has gained this position after attaining one of the highest densities of population in the world (Sreebha 2008).

Foreign remittances are the principal means by which emigration impacts the economy (Zachariah and Rajan 2004) and migration to Gulf countries continue to contribute to the state economy in a significant manner. Globalization has opened up a wider set of opportunities in the state and foreign remittances account for nearly a quarter (23 %) of the state income. All these notably enhanced the pace of economic growth in the state, leading to increased resource consumption and consequent environmental degradation. Added to these are the problems arising from the reduction in household size. The household number is now between 3 and 4 against its higher figures in early 1970s. Studies show that, even as the size of population remains constant, more household numbers imply a larger demand for

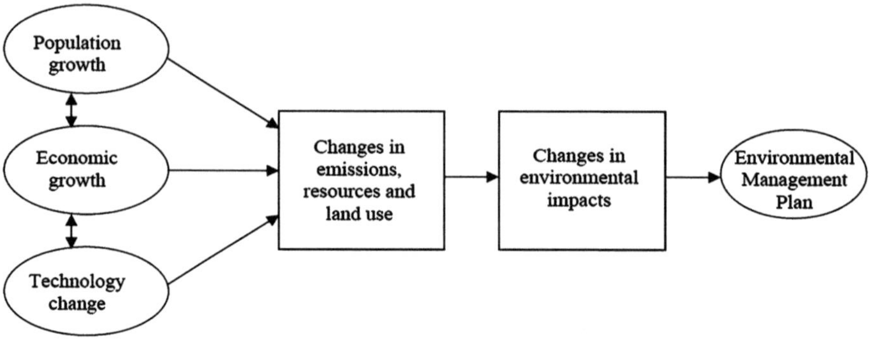

Fig. 6.15 Schematic representation showing the relation among three main drivers of environmental change (after Rubin 2001; Sreebha and Padmalal 2011)

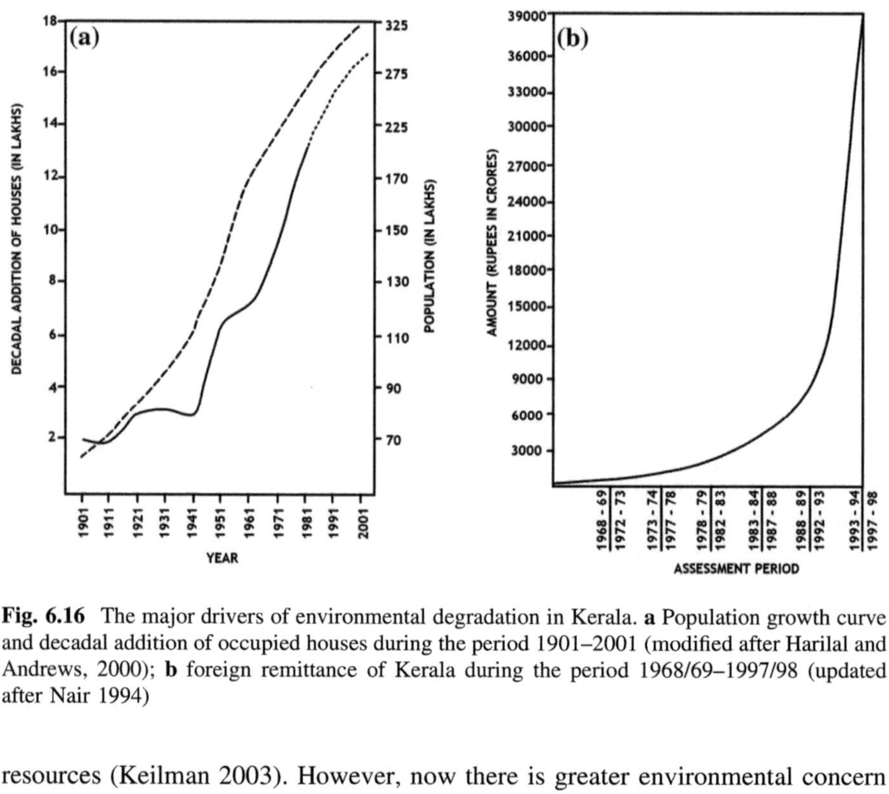

Fig. 6.16 The major drivers of environmental degradation in Kerala. **a** Population growth curve and decadal addition of occupied houses during the period 1901–2001 (modified after Harilal and Andrews, 2000); **b** foreign remittance of Kerala during the period 1968/69–1997/98 (updated after Nair 1994)

resources (Keilman 2003). However, now there is greater environmental concern among the people of Kerala, reiterating the fact that Kerala is at present at the helm of Kuznets curve (turning point) (Fig. 6.14) and there will be marked improvements in the quality of environment in the decades to come.

Table 6.4 Population density and decadal addition of persons in Kerala from 1901 to 2011

Sl. No.	Year	Population density (inh. km^{-2})	Decadal addition (inh. km^{-2})
1	1901	165	–
2	1911	184	19
3	1921	201	17
4	1931	245	44
5	1941	284	39
6	1951	349	65
7	1961	435	86
8	1971	549	114
9	1981	655	106
10	1991	749	94
11	2001	819	70
12	2011	859	40

6.5 Environmental Implications of Sand Mining

River channels are built up and maintained by erosional and depositional processes under different flow regimes. In any river, the velocity of water varies significantly along and across the river profile. The depth of the river has a direct bearing on its water flow/velocity. One of the important processes that enhance the depth of a river, other than natural processes, is the uncontrolled dredging of sand and gravel from its channel environment. As a result of sediment mining, the rivers of Kerala have experienced remarkable channel adjustments, particularly incision and narrowing in bar built areas (Plate 6.3) and channel widening in full flow channels. In the Manimala river, bed-level lowering has been of the order of 9–10 m at many locations whereas in the Periyar, bed has lowered to about 7.40 m during the period 1980–2010 (Padmalal et al. 2011). An intercomparison of the river bed lowering of the Kerala rivers with that of the published accounts of Rinaldi et al. (2005) and Lu et al. (2007) (Fig. 6.17) revealed that the rate of bed lowering of Kerala rivers is almost in agreement with that of the other small rivers of the world. In the case of Manimala river, incision and narrowing have produced significant changes in the channel pattern and morphology in the midlands and induced channel widening in lowlands. Besides morphological effects on river channels, sand mining in these two rivers have produced many adverse effects on engineering structures and also on their natural environmental settings. There has been significant lowering of groundwater table due to channel incision and an increase in pollution risk due to exposure of the water table in some mining pits. Creation of deep pits as well as ponding of water within river channels and floodplain areas, migration of pits and subsequent undermining of side protection walls, and other engineering structures are frequent in the storage zones of both Manimala and Periyar rivers. River bank slumping is noticed at many locations in the rivers especially in the midland and lowland physiographic units where river sand extraction is at an alarming rate compared to the rest of the river reaches.

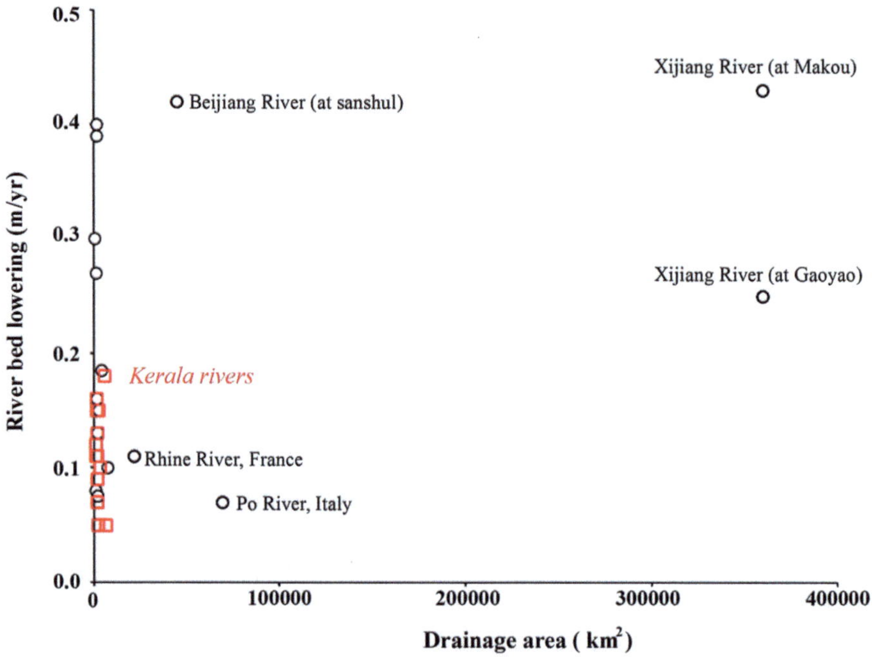

Fig. 6.17 Comparison of river bed lowering of rivers that hit indiscriminate sand mining (*Data source* Rinaldi et al. 2005; Lu et al. 2007 and Padmalal 2010)

Water scarcity problems are severe in the lowland and midland reaches of the river basins. A survey conducted as a part of this investigation revealed that the majority of the wells in the mining-hit local bodies of Periyar river is affected by water scarcity due to river bed lowering (channel incision) and subsequent lowering of the ground water table in the area. The analysis of CWC data reveals that the Periyar river bed at Neeleswaram gauging station has lowered to 392 cm with an average incision rate of 13.08 cm y^{-1} during the period 1980–2010. In addition, there are several reports of aggravated salt-water ingression consequent to mining induced channel incision, especially in areas close to river mouth zone which accentuates the water crisis situation further.

Apart from physical changes, the river ecosystems of Kerala are also facing a steady decline in the biological wealth, which is accentuated by caving of river banks and the inbuilt riparian flora. Sand mining enhances the level of suspended solids in the overlying water column, which in turn leads to higher turbidity levels and reduces light penetration and photosynthetic activity (Unni 2003; Harikumar et al. 2003). In Kerala rivers, habitat loss occurs mainly through indiscriminate scooping of sand and subsequent changes in the channel morphology and river bank erosion/slumping. Further, scooping of sand along with organic detritus and subsequent exposure of hard rocks/hard substratum may adversely affect the feeding, hiding, and breeding habitats of fishes and other aquatic organisms.

Plate 6.3 The eroding banks of Vamanapuram river in its downstream reach. **a** Palmyra palm (shown by the *arrow*) is located behind a *thick cover* of coconut groves; **b** river bank reaches to the location of the Palmyra palm; **c** within a short span of time (1999–2001), even the Palmyra palm has been uprooted by rampant scooping of construction grade sand from prohibited areas close to river bank and subsequent slumping/erosion during high flow regime of the monsoon period

Specific feeding habitats of species are hindered or interrupted which eventually threatens the biological diversity of the riverine systems.

River sand is a nonrenewable natural resource in terms of human life scale. If the quantity of sand extraction is within natural replenishments, the resulting environmental problems of sand mining will be minimal. But it is unfortunate that the indiscriminate extraction of river sand has changed the morphology of the rivers of Kerala to such levels that it will be difficult to reinstate their natural ecosystem functioning. The situation demands for high priority and specific policy interventions to restore the natural riverine character of the already disfigured rivers of the state. In this context, the role of government and communities are extremely important not only to ameliorate the situation but also to sustain both economic and environmental quality of river systems of the state.

References

Arun PR, Sreeja R, Sreebha S, Maya K, Padmalal D (2006) River sand mining and its impact on physical and biological environments of Kerala rivers, southwest coast of India. Eco-chronicle 1:1–6

CESS (1984) Resource atlas of Kerala. Centre for Earth Science Studies, Thiruvananthapuram

Census of India (2001) Provisional population totals. Registrar General and Commissioner of India, Ministry of Home Affairs, New Delhi, India

CWRDM (1995) Water atlas of Kerala. Centre for Water Resources Development and Management, Kozhikode, p 82

Douglas I (2005) Urban geomorphology. In: Fookes PG, Lee EM, Milligan G (eds) Geomorphology for engineers. CRC Press, Whittles Publishing, Boca Raton, pp 757–779

Eyles N (1997) Environmental geology of urban areas. Geological Association of Canada, St. John's, p 590

Gupta A (1984) Urban hydrology and sedimentation in the humid tropics. In: Costa JE, Fleisher PJ (eds) Developments and Applications of Geomorphology. Springer, Berlin, pp 240–267

Harikumar PS, Madhavan K, Sreedharan KE (2003) Water quality assessment of river basins of greater Kochi region. In: Abstract of international seminar on integrated river basin management with special reference to the humid tropics, CWRDM, Kozhikod, 22, 23 Aug 2003

Harilal KN, Andrews M (2000) Building and builders in Kerala: commodification of buildings and labour market dynamics. Discussion paper No. 22, KRPLLD, Centre for Development Studies, Thiruvananthapuram, p 30

Joy CM (1992) River Periyar and pollution problems. In: Michael P (ed) Environmental hazards in Kerala–problems and remedies

Kannan KP (2004) Kerala's challenges in a globalizing world. In: Basheer KM (ed) Memorial Lecture. Vakkom Moulavi Foundation Trust, Thiruvanathapuram, p 10 (Contributory paper)

Keilman N (2003) The threat of small households. Nature 421:530–533

Lu XX, Zhang SR, Xie SP, Ma PK (2007) Rapid channel incision of the lower Pearl river (China) since the 1990s as a consequence of sediment depletion. Hydrol Earth Syst Sci 11:1897–1906

Milliman JD, Syvitski JPM (1992) Geomorphic/tectonic control of sediment discharge to the ocean: the importance of small mountainous rivers. J Geol 100:525–544

Nair PRG (1994) Migration of Keralites to the Arab world. In: Kerala's economy—Performance, problems and prospects. Sage Publications, New Delhi

Padmalal D, Maya K, Sreebha S, Sreeja R (2008) Environmental effects of river sand mining: a case from the river catchments of Vembanad lake, southwest coast of India. Environ Geol 24:879–889

Padmalal D (2010). Alluvial sand mining: the Kerala experience. Commissioned papers—Western Ghats. http://westernghatsindia.org/commssioned-papers

Padmalal D, Maya K, Baijulal B (2011) Environmental appraisal and sand auditing of Periyar river, Ernakulam district, Kerala. Centre for Earth Science Studies, Thiruvananthapuram, pp 1–212

PWD (1974) Water resources of Kerala. Government of Kerala, India

Rao MBR (1973) The subsurface geology of the Indo-Gangetic plains. J Geol Soc India 14:217–242

Rinaldi M, Simoncini C, Sogni D (2005) Variazioni morfologiche recenti di due alvei ghiaiosi appenninici: il F. Trebbia ed. il F. Vara. Geografia Fisica e Dinamica Quaternaria Suppl 7:313–319

Rubin ES (2001) Environmental forecasting. In: Rubin ES (ed) Introduction to engineering and the environment. Water resources and environmental engineering series, McGraw Hill, New York, pp 635–677

Shiekha EJ, Maya K, Padmalal D (2014) Environmental effects of river sand mining: a case study of Periyar river, Kerala, SW India. In: Shaji E, Pradeepkumar Ap (eds) Mineral Resources of Kerala. University of Kerala, India, pp 115–125

Sreebha S (2008) Environmental impact of sand mining: a case study in the river catchments of Vembanad lake, Southwest India. Ph.D. Thesis (Unpublished), Cochin University of Science and Technology, Kochi

Sreebha S, Padmalal D (2011) Environmental impact assessment of sand mining from the small catchment river in the southwestern coast of India. Environ Manage 47:130–140

Unni KS (2003) Biodiversity in the river ecosystems of Peninsular India with special reference to Kerala. In: Abstract on international seminar on integrated river basin management with special reference to the humid tropics. CWRDM, Kozhikod, 22, 23 Aug 2003

Zachariah KC, Rajan SI (2004) Gulf revisited economic consequences of emigration from Kerala. Emigration and unemployment, Working paper 363, Centre for Development Studies, Thiruvananthapuram, p 88

Chapter 7
EIA of River Sand Mining

Abstract Depending on the geological and geomorphological setting, the magnitude of the environmental impacts of river sand mining also varies. The activity imposes marked environmental impacts on land, water, atmosphere, and socio-economic conditions of a river basin. Environmental impact Assessment (EIA) is an important tool to assess the positive (beneficial) and negative (adverse) impacts of the sand mining. Two different methods—Matrix method and Rapid Impact Assessment Matrix method are used here to assess the environmental impact of sand mining in the longest river of Kerala—the Periyar river. The study reveals that the activity is a major threat to the very existence of the Periyar river in particular and the small rivers of tropics in general. The rivers are reported to be degraded considerably due to uncontrolled sand mining over the past few decades. The impact assessment reveals that the adverse environmental impacts of sand mining dominate over the marginal and short-term benefits.

Keywords Environmental impact assessment of sand mining · Matrix method · Rapid impact assessment matrix (RIAM) · Environmental management plans · River restoration

7.1 Introduction

There can never be any doubt about the role played by the mineral sector in economic development. Extractive industries of major and minor (building materials) minerals have been the backbone of development processes. At the same time, it should be remembered that we do not have infinite quantity of resources and their finite supply can be depleted within the human life scale. Also

D. Padmalal and K. Maya, *Sand Mining*, Environmental Science and Engineering,
DOI: 10.1007/978-94-017-9144-1_7,

important is, the environmental problems that could be surfaced from large-scale mining of natural resources of economic significance. Therefore, the question to be addressed with regard to environmental problems of mining pertains to conservation of minerals and the respective ecosystem(s) on one hand and environmental management activities/restoration on the other.

Depending on the location and type of mining, the magnitude and importance of environmental impacts would also vary. These impacts are broadly categorized under four categories—impact on land, atmosphere, water and socioeconomic condition of the people of the region (Rau and Wooten 1980). The main issues that are to be addressed while dealing with the environmental problems of mining activities including sand and gravel extraction are as follows:

(i) Water pollution
(ii) Air and noise pollution
(iii) Land degradation
(iv) Landslide and land stability
(v) Deforestation
(vi) Disruption of water regimes and drainage channels
(vii) Human environmental problems (settlement, health, employment and allied problems), and
(viii) Damages to sites of topographic, cultural, historical and scenic importance.

All these problems are to be investigated carefully while arriving at conclusions and also for formulating management strategies for mining-related environmental problems.

7.2 Environmental Impact Assessment

The sand and gravel mining has become one of the largest mining sectors in southwest India, especially in Kerala state as the region is fast growing due to economic developments in the past 4–5 decades. Large quantities of sand and gravel are extracted from the active channels and floodplain areas of Kerala rivers which constitute a convenient source of fine aggregates for the ever increasing construction sector (Padmalal et al. 2008). Considering the severity of the effects of sand mining on various environmental components of river ecosystems, an EIA study was undertaken to assess the impact of sand mining in one of the important rivers of Kerala—the Periyar river as an example. The environmental setting of the river has already been depicted in the previous chapter.

Environmental Impact Assessment is an important tool to select the best option among a set of alternatives having different degrees of adverse environmental impacts. Though the conventional EIA is of predictive nature, the present assessment of river sand mining is a *de facto* record of impacts, which has already occurred in mining-hit areas and its sub-environments as a result of the sand extraction processes. This is a added advantage as it provides information to the decision makers as well as the general public about the real picture of the environmental impacts incidental to river sand mining.

7.3 General Methodologies of EIA: An Overview

Various methodologies are available for the assessment of adversities of an activity on the naturally evolved environment/ecosystems. Some of the important methodologies are described briefly in the following sections.

7.3.1 Ad hoc Method

It is a simple method that merely presents the impacts without compaction. A good example of ad hoc method is when a team of experts assemble for a short time to conduct an EIA. Each expert's conclusions are based on unique combination of experience, training, and intuition. These conclusions are assembled in the form of a report. It gives no assurance about adversities of all the impacts caused by the activity in the natural environment.

7.3.2 Checklist Method

It is a comprehensive list of environmental effects and impacts which would enable the analyst to think broadly, and to predict the possible impacts of various actions. Checklists can be designed in several forms, which make it a very popular procedure for conveying EIA. A major drawback attributed to this method is that an analyst is likely to ignore factors that are not normally enlisted. For a short-term EIA made at a lower cost, this method is very useful. In the checklist method, the different elements of the environment are listed on the left hand side of the matrix and the impacting actions are listed across the top of the matrix. The entries in the matrices are defined in terms of"positive impacts" (favorably improving the environment) and "negative impacts" (defined as disrupting or otherwise adversely affecting the environment or services).

7.3.3 Matrix Method

The environment impact of a project or an action encompasses a broad range of impacts on land, air, vegetation and wildlife, socioeconomic aspects,esthetics, etc. These impacts vary both in magnitude and importance; may be beneficial of or adverse in nature. Because of this complex dimension of various impacts, a 'collective' or 'overall' environmental assessment of the project or action also becomes necessary. Several types of "ranking," "rating," or "weighting" of impacts are available in the matrix methodology.

7.3.4 Networks

It identifies direct and indirect order of impacts. The method also identifies and incorporates mitigation and management strategies into planning stages of a project. Though it is suitable for expressing ecological impacts, it has lesser potential for considering social, human, and esthetic aspects.

7.3.5 Overlays

This system is useful while addressing questions at site and in route selection. It provides an effective mode of presentation and display to the audience. There is no provision for quantification and measurement of impacts. Overlay analysis are purely spatial where social, human, and economic aspects are not covered.

7.3.6 Simulation and Modeling Workshop

A simplified treatment of different models used in EIA has been furnished by Munn (1975). It provides enormous scope for understanding the behavior of environment vis-a-vis impacting actions. However, the extreme complexities involved in geological and sociological factors make modeling equally difficult as it is with other methods. The most practical model that is presently available is mathematical modeling, in which the physical connections are replaced by logical relationships. The mathematical simulation provides greater flexibility, especially when a computer is available to solve the mathematical equations. It is possible to investigate the consequence of many options and in this possibility lies the particular advantage of computer modeling in EIA.

7.3.7 Rapid Impact Assessment Matrix

This method seeks to overcome the problems of recording subjective judgments by defining the criteria and scales against which these judgments are to be made and by placing the results in a simple matrix that gives a permanent record in the judgment process. EIA has progressed from the consideration of pollution assessment, though the wider range of ecological assessment and has now become holistic in nature (Pestakia 1998). Rapid Impact Assessment Matrix (RIAM) is based on a standard definition of the important assessment criteria, as well as the means by which semi-quantitative values for each of these criteria can be collected to provide a fairly accurate and independent score for each condition.

7.4 Methodology Adopted

In the EIA of the sand and gravel extraction from river environments two important EIA methodologies are adopted—(1) the matrix method prescribed by Rau and Wooten (1980) and (2) Rapid Impact Assessment Matrix (RIAM) proposed by Pastakia (1998). These two methods are applied for a single activity, i.e., sand and gravel extraction, for a better understanding of the activity on one hand and to evaluate the methods on the other.

The matrix method is basically a method for presenting various impacts in an abstract way. It starts with a list of potential impact areas, which are normally environmental conditions or characteristics, and the nature of impacts as to whether negative (adverse) or positive (beneficial). These are assessed against various impacting actions. In this approach, the environmental conditions/characters are listed on the left hand side of the matrix and the impacting actions on the top.

In RIAM, initially the impacts of the project activities are evaluated against the environmental components/subcomponents. Then for each individual component a score is assigned, which provides a measure of the impact expected for the component (Pastakia 1998). The important assessment criteria fall in two groups: (1) Group A—criteria that are important to conditions and, (2) Group B—criteria that are of value to the situation. The values allotted to each of these groups of criteria are determined following Pastakia (1998).

7.5 EIA of Sand and Gravel Mining

7.5.1 Impact Assessment Using Matrix Method

Sand extraction from river channels could cause local disruptions in the river environment. After recording various environmental problems of river sand mining in different physiographic zones, an EIA was carried out to suggest appropriate Environmental Management Plan (EMP) for regulating the mining activities on sustainable basis. The river environments in all the three physiographic zones such as highlands, midlands and lowlands are considerably deteriorated due to the illicit scooping of sand even from prohibited areas close to bridges and water intake structures. Hence, an attempt has been made to analyze the environmental impacts caused by river sand mining in order to identify and address the key environmental issues that are resulted from the activity. The main purpose of the effort is to mitigate the negative impacts and enhance the positive ones.

The magnitude of impacts arising from instream sand mining in the storage zones in the midlands and lowlands of the study area are shown in Table 7.1. Impacting actions are those that significantly affect the quality of the environment in which the overall cumulative primary and secondary impacts alter the quality of the environment. The threshold of the magnitude and importance of impacts is implied by the terms "low," "medium," and "high" and must be met before an Environmental Impact Statement (EIS) is prepared. The important impacting actions which have been considered here are the ones that have either "high" positive or "high" negative impacts on the environment. This, in no way, belittles the significance of lesser impacts. Cases of 'no appreciable impacts' or 'undetermined impacts' are also represented. The mode of mining is manual in the highlands and midlands. But in lowlands, in addition to manual mining, mechanical/semi mechanical mining is also reported by the local people. Bar skimming and pit excavation are the commonly adopted methods of sand extraction in the highlands and midlands. But in lowlands, the widely practiced method is pit excavation. The extracted sand from the midlands and lowlands require cleaning as it contains clay lumps, decayed wood and, in some cases, shells, whereas in highlands, material cleaning is not required. Instead, in certain cases, the bed materials have to be graded using frame screens of different mesh sizes to recover the desirable size grades.

The impacts of sand and gravel mining from rivers and floodplains affect many environmental subcomponents in the area, including land and water, biota, public health, and safety. The database generated from the study and the conclusions drawn from it could be utilized for chalking out strategies for the conservation and sustainable management of resources. It can also be used for framing guidelines for regulating the mining activities in an environment-friendly basis. The EIA performed here depicts the growing awareness and increasing public concerns over the impacts of river sand mining on the biophysical environment of riverine system.

Table 7.1 Environmental Impact Assessment (EIA) of instream sand mining in the study area using matrix method

Components of environment			Impacting actions							
			Mining				Post-mining			
Environmental conditions / Parameters		Sub-components	Mode of Mining		Type of mining		Processing		Transportation	
			MaE	MeE	BS	PE	SG	MC	AR	VM
Physical & Chemical	Land / River channel	Land stability	●	⬤	•	⬤	✡	✡	•	●
		Landuse / Land cover	•	●	✡	●	•	✡	•	•
		Soil	●	⬤	•	⬤	•	✡	•	●
		Landform	●	●	•	•	✡	✡	✡	✡
		River bed	⬤	⬤	•	⬤	✡	•	✡	✡
	Air	Air quality	✡	•	✡	✡	✡	✡	✡	●
		Noise level	✡	•	✡	✡	◇	✡	✡	●
	Water	Groundwater	●	⬤	✡	⬤	✡	✡	✡	✡
		Surface water	⬤	⬤	•	⬤	✡	●	✡	•
Biological	Flora & fauna	Terrestrial	●	●	●	●	•	✡	•	•
		Riparian	●	⬤	•	⬤	•	✡	•	●
		Instream	⬤	⬤	●	⬤	✡	●	•	✡
Social	Socio-economic	Employment	⬛	■	■	⬛	■	■	▪	■
		Economic base & construction sector	⬛	⬛	■	⬛	■	▪	▪	■
		Land values and holdings	⬤	⬤	●	⬤	•	•	⬛	✡
		Engineering structures	⬤	⬤	•	⬤	✡	✡	✡	●
	Socio-health	Accidents & Health impairment	⬤	●	•	⬤	✡	✡	✡	●
	Socio-cultural	Sand bars	•	●	•	●	✡	✡	✡	✡
	Socio-livelihood	Sustainable livelihoods (Fishing, farming etc.)	⬤	⬤	•	⬤	✡	◇	✡	✡

• Low negative impact ● Medium negative impact ⬤ High negative impact ▪ Low positive impact ■ Medium positive impact
⬛ High positive impact ✡ No appreciable impact ◇ Undetermined impact MaE: Manual extraction MeE: Mechanical extraction
BS: Bar skimming PE: Pit excavation SG: Size grading MC: Material cleaning AR: Approach road VM: Vehicular movement

7.5.2 Rapid Impact Assessment Matrix

7.5.2.1 Assessment Criteria

The assessment criteria used under RIAM includes two groups.

Group A: Criteria that are of importance to the condition, and which can individually change the score obtained.

Group B: Criteria that are of value to the situation but should not be individually capable of changing the score obtained.

The value allotted to each of these groups of criteria is determined by the use of a series of simple formulas as illustrated by Pastakia (1998). It allows the scores of the individual components to be determined on a definite basis. The scoring system requires simple multiplication of the scores given to each of the criteria in Group A (A1 & A2). The use of multiplier for Group A is important as it immediately ensures that the weight of each score is expressed, whereas simple summation of scores could provide identical results for different conditions. Score for the value criteria B are added together to provide a single sum. This ensures that the individual value scores cannot influence the overall score, but the collective importance of all values in Group B (B1, B2 & B3) are fully taken into account. The sum of the Group B scores is then multiplied by the result of the Group A scores to provide a final assessment score, ES for the condition.

The process can be expressed by the formula,

$$\text{If } (a_1) * (a_2) = \text{aT}$$

and

$$(b_1) + (b_2) + (b_3 = \text{bT})$$

Then $(\text{aT}) * (\text{bT}) = \text{ES}$

Where,

a_1 and a_2 are the individual criteria scores for Group A,
b_1 to b_3 are the individual criteria scores for Group B,
aT is the result of multiplication of all A scores,
bT is the result of summation of all B scores,
ES is the environmental score for the condition.

Positive and negative impacts can be demonstrated by using scales that pass from negative to positive values through zero for the Group A criteria. Zero thus becomes the no change or no importance value. The use of zero in this way in Group A criteria allows a single criterion to isolate conditions which show no change or are important to the analysis. Zero is a value avoided in the Group B criteria. If all Group B criteria score zero, the final result of the ES will also be zero. This condition may occur even where the group A criteria show a condition of importance that should be recognized. To avoid these scales for the Group B criteria use 1 as the no change or no importance score. The criteria should be defined for both groups, and should be based on fundamental conditions that may be affected by change rather than be related to individual projects. It is theoretically possible to define a number of criteria, but two principles should always be satisfied:

1. The universality of the criterion, to allow it to be used in different EIAs
2. The value of the criterion, which determines whether it should be treated as Group A or Group B condition.

At this point, only five criteria have been developed for use in the RIAM. Nevertheless these five criteria represent the most important fundamental assessment conditions for all EIAs, and satisfy the principles set out above.

7.5.2.2 Assessment Results

RIAM makes it possible to carry out an analysis of the results based on individual Environmental scores (ES) for each environmental components/sub-components that are classified in ranges so that the effects can be compared to each other. The

Table 7.2 Environmental components/subcomponents and impact categories of sand mining from Periyar river (after Pastakia 1998; Resmi et al. 2011)

Components of environment			Assessment criteria					ES	RV
Environmental conditions/parameters		Subcomponents	Group A		Group B				
			A1	A2	B1	B2	B3		
Physical and chemical	Land/river channel	Land stability	1	−2	3	3	3	−18	−B
		Landuse/landcover	1	−2	2	2	3	−14	−B
		Soil	1	−2	2	2	3	−14	−B
		Landform	1	−3	3	3	3	−27	−C
		River bed	1	−3	3	3	3	−27	−C
	Air	Air quality	1	−1	1	1	1	−3	−A
		Noise level	1	−1	1	1	1	−3	−A
	Water	Ground water	1	−3	3	3	3	−27	−C
		Surface water	1	−2	2	2	3	−14	−B
Biological and ecological	Flora	Instream flora	1	−2	2	2	3	−14	−B
		Riparian flora	1	−2	2	2	2	−12	−B
	Fauna	Instream fauna	1	−2	2	2	2	−12	−B
		Riparian fauna	1	−2	2	2	2	−12	−B
	Habitat	Habitat loss	1	−2	2	2	3	−14	−B
Social and cultural components	Social-health	Accidents	1	−1	1	1	1	−3	−A
		Health impairment	1	−1	1	1	1	−3	−A
	Social-cultural	Heritage/historical areas	1	−3	3	3	2	−24	−C
	Socio-livelihood	Sustainable livelihoods (Fishing, farming etc.)	1	−2	2	2	2	−12	−B

(continued)

Table 7.2 (continued)

Components of environment			Assessment criteria					ES	RV
Environmental conditions/parameters		Subcomponents	Group A		Group B				
			A1	A2	B1	B2	B3		
Economic and operational environment	Economic	Employment	1	2	2	2	2	12	B
		Economic base	1	2	2	2	2	12	B
		Agriculture	1	−2	3	3	3	−18	−B
		Aesthetics	1	−2	2	2	2	−12	−B
	Operational	Approach road	1	−2	3	3	3	−18	−B
		Engineering structure	2	−3	3	2	3	−48	−D
		Infrastructure	2	3	3	3	3	54	D
		Transportation	2	−3	3	3	3	−54	−D

Note: *A1* Importance of condition (International importance/ scale 4, National importance/ 3, Outside of local condition/ 2, Local condition/ 1, not important/ 0); *A2* Magnitude of change or effect (Major positive benefit/ scale +3, Significant/ +2, Improvement/ +1, No change to status quo/ 0, Negative change to status quo/ -1, Significant negative impact/ -2, Major negative impact/ -3); *B1* Importance of performance (No change or no effect/ score 1, Temporary/ 2, permanent/ 3); *B2* Reversibility (No change or no effect/ scale 1, Temporary/ 2, Permanent/ 3); *B3* Cumulative (No change or no effect (scale 1, Non-cumulative or single/ 2, Synergistic/ 3); *ES* Environmental score; *RV* Range value.

Table 7.3 Summary of assessment of RIAM of sand mining from Periyar river (after Yousefi et al. 2009; Resmi et al. 2011)

Si no.	ES	Range value		Description	Environmental components					Final[a]	Impact total (in %)
		Alphabetic	Numeric		PC	BE	SC	EO	Total		
1	72–108	E	5	Major positive change	0	0	0	0	0	0	0
2	36–71	D	4	Significant positive change	0	0	0	1	1	4	7
3	19–35	C	3	Moderate positive change	0	0	0	0	0	0	0
4	10–18	B	2	Positive change	0	0	0	2	2	4	7
5	1–9	A	1	Slight positive change	0	0	0	0	0	0	0
6	0	N	0	No change/status quo	0	0	0	0	0	0	0
7	−1–−9	−A	−1	Slight negative change	2	0	2	0	4	−4	7
8	−10–−18	−B	−2	Negative change	4	5	1	3	13	−26	45
9	−19–35	−C	−3	Moderate negative change	3	0	1	0	4	−12	20
10	−36–−71	−D	−4	Significant negative change	0	0	0	2	2	−8	14
11	−72–108	−E	−5	Major negative change	0	0	0	0	0	0	0

ES Environmental score; *PC* Physical and chemical; *BE* Biological and Ecological; *SC* Social and Cultural; *EO* Economic and operational

[a] Product of range values (numerical) and environmental component's total

description of the components and the impact categories in the assessment process are depicted in Table 7.2. Table 7.3 summarizes the final results of the RIAM process. From the tables, it is evident that most of the impacts are in the negative end and the benefits the activity generates are very limited and are rather a few short-term economic gains. Considering the magnitude of the negative impacts, strict measures are required to rescue Periyar river from the uncontrolled sand mining operations which is widespread all along the river channel.

7.5.3 Physical and Chemical Components

Mining is the process of extraction of non-living resources from the earth's surface. The activity in the long run adversely affects the physical as well as chemical characteristics of river environment. The accuracy with which one can assess physical impacts is dependent on the understanding of the functions of natural environment. Perfect understanding is necessary to predict exactly the direction, degree and rate of change of the physical environment. Depending on the type and extent of mining, the physical effects of dredging process depend on and varies significantly. Physical effects of indiscriminate sand mining like river bed lowering, bank slumping, channel instability, etc., may end up in the total degradation of the riparian and aquatic ecology, in addition to undermining engineering structures. The specific environmental issues of river sand extraction in the Periyar river is addressed in detail in this Chap. 6.

7.5.3.1 Land/River Channel

The manual mining practiced in the extraction of sand and gravel could induce negative changes on land stability, soil, landform, esthetics, and riverbed characteristics. The stability of the riverbank is seriously affected by intense sand mining activities as the banks are made up of incompact soil. Invariably, the fluvial landforms (alluvial/sand bars) within the river channel are removed from the storage zones. The secondary effects of sand mining, viz., river bank slumping, channel incision, etc., cause obvious physical disturbances on the land stability especially in areas close to the riverbanks. Manual mining causes high negative impacts on the riverbed. Mechanical mining could impose high negative impact on the stability of land and soil close to the riverbanks, and also on river bed, in addition to imposing marked changes (medium negative impacts) on land use, landform and aesthetics of the area. Due to indiscriminate sand mining, pits of various dimensions are formed on the riverbeds. Formation of deep pits in the channel bed produce slower flow velocities and lower flow energies, causing sediments transported from upstream to deposit in mining pits. The resultant sediment deficient water (hungry water) can have a very high erosive effect on the river banks downstream triggering bank erosion and damages to engineering structures. There are also downstream impacts arising from increased turbidity and sedimentation affecting the quality of water. It is quite evident that any type of mining process would drastically change the existing aesthetic environment and adversely affect the natural environment. Processing activities like size grading and material cleaning creates a low negative impact on aesthetics.

7.5.3.2 Air and Water

The impacting actions like manual mining would not produce much direct effect on air quality. However, the mining-related activities like movement of vehicles through unplanned/non-paved roads produce marked adverse effects on the ambient air quality, in addition to enhancing the noise level. Mechanical mining in the lowland areas can cause negative impacts on air quality and noise, though at lower levels (Table 7.1). The frequent movement of haulage trucks leads to compaction of soil which, in turn, adversely affects the net productivity of the land.

The after effects of river sand mining such as siltation turbidity, creation of deep pits in the channel, transportation of vehicles into the riverbed for easy loading of the mined material, spillage of oil/gasoline from the vehicles, etc., adversely affect the surface water quality. Bar skimming and pit excavation in the highlands could create low–medium negative impacts in river water. In highlands, there won't be any appreciable level of impacts on groundwater resources due to sand mining, whereas midlands and lowlands are the most vulnerable to sand mining. Manual mining in storage zones have medium negative impact on groundwater resources and high negative impact on river water. Among the two types of mining, pit excavation is more harmful as it adversely affects the environmental setting of the midlands and lowlands. The enhanced level of suspended solids smothers the benthic community, clogs the gills of fishes and is a serious threat to the existence of aquatic biota Turbidity reduces vision and mask odors, both important for the survival of many fishes. The conditioning and grading (or processing) of raw sand can aggravate the suspended sediment contents in the overlying waters.

7.5.4 Biological Components

Of the different modes of mining, manual mining causes medium negative impact on terrestrial fauna. At the same time the riparian flora and fauna in the highlands experience low negative impacts. The instream biota is the most affected by the manual mining processes. Among the different types of mining, bar skimming has medium negative impact on terrestrial and instream biota. Indiscriminate mining from the sand bars during the past few decades has lead to the destruction of resting and often nesting grounds of many migratory birds. Natural riparian habitats are characterized by variable radiant of moisture and light, lush vegetation, and very high biodiversity. Lowering of water table due to indiscriminate sand mining can induce loss of riparian vegetation along the banks of rivers, i.e., if the water table drops below the root zone. Riparian flora and fauna suffer from serious effects caused by river bank slumping, channel incision, lowering of water table, etc., as a result of the direct removal of vegetation along the river banks, bank undercutting and channel incision (Brown et al. 1992; Price and Lovett 2002a, b; Baijulal et al. 2011). The deep pools shaded by the over hanging riparian shrubs and trees, are vital resting sites for instream fauna, especially fishes. Since aquatic ecosystems are among the most productive and diverse habitats, it is not surprising that they are easily upset by a variety of perturbations caused by human beings. The river itself has unique bottom characteristics, temperature, velocities, bank features, riparian vegetation, and invertebrates that serve as fish food. Spawning must occur in the gravel/sand beds which act as incubators where eggs are well oxygenated and are hidden from predators. The newly hatched fishes escape from the gravel beds and seek out the stream nurseries. The gravel beds are the shallow impoundments of quiet water which is characterized by warmer temperature, protective vegetation cover, and abundant invertebrates (for food) all of which supports the spawning habitat. Thus the distribution of instream biota is strongly related to the physical habitat (Brown et al. 1998). The various adverse impacts of river sand mining easily disrupt the complex relationship between physical and biological factors of aquatic systems.

7.5.5 Social Components

Sand is an important aggregate in the construction sector. Therefore, the mining of the same from rivers and adjoining environments causes a major positive impact in the socioeconomic sector. Through mining activities, jobs and opportunities are created, and significant contributions are made to the State's economy. It is also an important source of revenue. However, these benefits are at a cost of direct and indirect impacts on the environment. The economic and social benefits of resource extraction, if any, are always for a short term, as the revenue ceases with the exhaustion of the resource. Activities which provide socioeconomic gains from the use of aggregate resources often result in the impairment of ecosystem functioning. The negative impacts of river sand mining often overweigh its short-term positive phase. Loss of human lives due to accidents has been rising over the years as the pits, partially filled with silt and clay deposits, act as death traps. Channel incision causes undermining of bridges, piers and other infrastructural facilities associated with river channels.

The magnitude of the impact varies according to physiographic zones and intensity of mining. Added to these is the impact of transportation of heavy vehicles for haulage of sand to the construction sites. The health of the people in the lowlands is also significantly affected by manual mining of sand. Depletion of sand in point bars and bars of Periyar river is another pressing problem the State faces. Sand bars in the river channel that are used for holding annual religious/ cultural congregations are degrading fast due to indiscriminate sand mining.

7.6 Discussion

Understanding fully the various dimensions of a river basin environment and its application to the conservation and management of the pristine freshwater systems is a daunting challenge to mankind. Management of rivers require legal, social and economic considerations, as well as scientific insight. Political will and development of institutional infrastructure enables management of rivers to keep pace with the rapid growth in science is critical for the conservation and restoration of this key ecosystem.

Rivers and their floodplains enfold diverse and valuable ecosystems. In the river ecosystem, the biotic community are closely interrelated. Fishes prey on the plants and insects, and are later themselves eaten by birds, amphibians, reptiles, and mammals. Apart from the river channel, wetlands that are maintained by seepage and is occasionally flooded by the river, supports rich and diverse habitats that are important for resident species, migrating birds and animals that use wetlands as staging posts while moving seasonally (domiciles); (Sreebha 2008). Riparian ecosystem is one of the key systems that depend entirely on river regime for their existence. Therefore, river conservation and management require a balanced/ holistic approach. Great care must be exercised while altering this system because any amount of careless handling or over-exploitation of resources (both living and non-living) may end up in catastrophic impacts on this unique ecosystem.

Adequate legislations are to be made to achieve these goals. Strategies used to manage instream mining vary widely, and in many areas there are no effective management strategies in practice. The ill-planned and careless extraction of sand and gravel from river channels have adverse effects on the sand bar stability and channel evolution. Impacts on fish habitat, aesthetics, and other river characteristics are indirect effects of the channel disturbances. Periodic field checking of river sand mining sites, record of the quantity sand extraction, environmental auditing, etc., could be used for formulation of appropriate strategies for the wise management of sand mining from river environments. Another approach for managing sand mining is to estimate the annual bedload supply from upstream, (i.e., the replenishment rate) and to limit annual extraction to that value or some fraction thereof (i.e., the safe yield). Here also periodic sand auditing is a must for evaluating the performance of the mining process.

Indiscriminate river sand and gravel mining over the years have imposed irreparable damages to river ecosystems. Lack of adequate scientific information on the mining processes and its impacts on various environmental components is a major lacuna for properly addressing the problem. Therefore, there is an imminent need for strengthening our understanding on anthropogenic degradation of the small catchment rivers. This is vital not only for laying down strategies for regulating the mining activities on an environment-friendly basis, but also for creating awareness on the impact of river sand mining on the physicochemical and biological environment of these life support systems. This study discloses the fact that the Periyar river has degraded considerably due to uncontrolled sand mining over the past few decades and the adverse impact of mining dominate at significant levels over the marginal and short-term benefits. This points to the urgent need for stringent measures against indiscriminate sand mining activity for improving the river health. A modified/expanded version of the 'redline concept' of Kondolf (1994) will be best for the estimation of mineable sand in the small rivers of Kerala in general and the Periyar river in particular.

References

Baijulal B, Jissy Jyothi S, Padmalal D, Maya K (2011) Environmental effects of alluvial sand mining on the biological environment of Kerala rivers. In: Proceedings of the national seminar on "mining of river sand and its impact on the environment", CWRDM, Kozhikode 18–19 Feb, p 87–110

Brown AV, Lyttle MM, Brown KB (1992) Impacts of gravel mining on Ozark stream ecosystems. Arkansas Cooperative Fish and Wildlife Research Unit, Fayetteville, p 118

Brown AV, Lyttle MM, Brown KB (1998) Impacts of gravel mining on gravel bed streams. Trans Am Fish Soc 127:979–992

Kondolf GM (1994) Environmental planning in the regulation and management of instream gravel mining in California. Landscape Urban Plan 29:185–199

Munn RE (1975) Environmental impact assessment: principles and procedures. Scope 5:160

Padmalal D, Maya K, Sreebha S, Sreeja R (2008) Environmental effects of river sand mining: a case from the river catchments of Vembanad lake, Southwest coast of India. Environ Geol 24:879–889

Pastakia MRC (1998) Rapid impact assessment matrix in environmental impact assessment RIAM. Olsen & Olsen, Fredensborg, p 8–19

Price P, Lovett S (2002a) Managing riparian land, fact sheet 1. Canberra, Land and Water Australia, p 8

Price P, Lovett S (2002b) Stream bank stability, fact sheet 2. Canberra, Land and Water Australia, p 13

Rau JG, Wooten DC (1980) Environmental impact analysis handbook. Mc Graw Hill Book Co., New Delhi, p 385

Resmi R, Baburaj B, Santhosh V, Sreebha S, Padmalal D (2011) EIA of river sand mining using rapid impact assessment matix (RIAM)—a case study. In: Proceedings of the national seminar on mining of river sand and its impacts on the environment, CWRDm, Kozhikkode, pp 183–189

Sreebha S (2008) Environmental impact of sand mining: a case study in the river catchments of Vembanad lake, Southwest India. Ph.D. thesis (Unpubl.), Cochin University of Science and Technology, Kochi

Yousefi H, Ehara S, Yousefi A, Seiedi F (2009) Environmental impact assessment of Sabalan geothermal power plant, NW Iran. In: Proceedings in 34th workshop on geothermal reservoir engineering, Stanford University, California

Chapter 8
Mining Strategies and Management

Abstract The exponential rise in the demand for construction grade sand resulted from economic growth and the liberal development policies over the years have aggravated indiscriminate scooping of sand and gravel from river beds and floodplains of most rivers in the world. The situation warrants scientific interventions for judicious management of the river bed resources by striking a balance between its demand for development and the environmental concerns. The present chapter reviews the existing regulatory procedures, and also environmental laws/ policies that are in practice in different parts of the world. A set of guidelines have been put forth to update the existing sand and gravel extraction policies in India and elsewhere for achieving the goal of sustainability in the case of sand resource extraction and managements.

Keywords Mining strategies · Environmental laws of sand mining · Sediment budgeting · Redline concept

8.1 Introduction

The economic growth and liberal development policies of nations have resulted in an exponential rise in construction sector. This has increased the demand for aggregate materials including the fine aggregates—sand and gravel. The indiscriminate extraction of sand and gravel from river and related aquatic environments have imposed severe impairments on these systems. The situation warrants scientific interventions for judicious management of the resources by striking a balance between the need for development and environmental problems (Ramachandran and Padmalal 1997) arising out of the mining and/or quarrying activities. Sustainable resource extraction needs to be achieved by every nation so that the natural resource requirements in future can also be catered to while satisfying the needs of the present generations. In view of the increased awareness of environmental problems, the accent on sustainable development has grown in recent

D. Padmalal and K. Maya, *Sand Mining*, Environmental Science and Engineering,
DOI: 10.1007/978-94-017-9144-1_8,

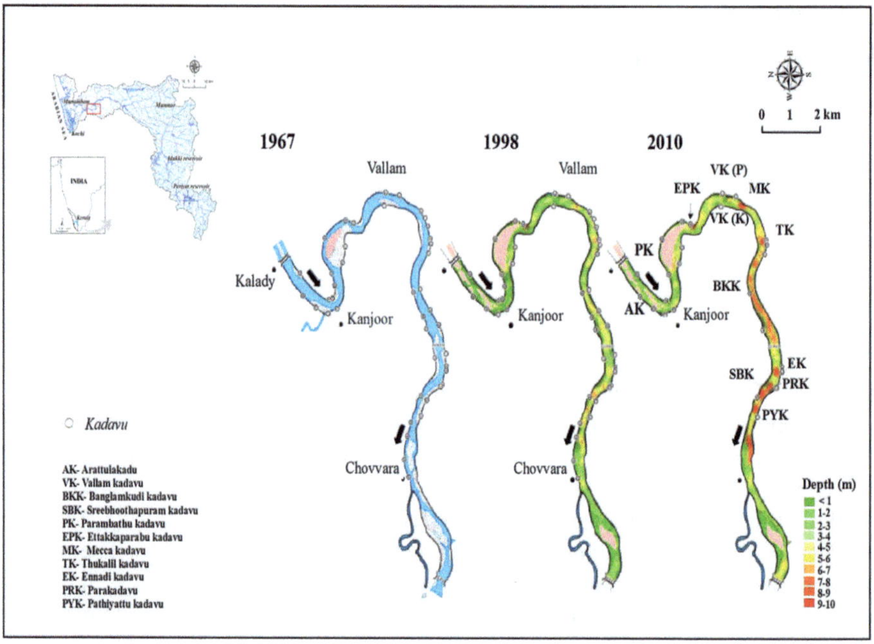

Fig. 8.1 Drainage map of Periyar river showing river depth changes in certain river stretches. Depth categorization is not attempted for the year 1967 as there is no bathymetric information available for the year

times, particularly with respect to activities, which degrade the environment and affect communities adversely. The after effects of exploitation of river systems are now becoming blatantly evident with the increasing problems of pollution hazards and channel instability. Of all the disruptive human interferences in river systems, river sand extraction is of special significance as it magnifies adversities like channel instability, channel morphology changes, river bed and bank erosion, river buffer zone encroachment, river water quality deterioration, etc. The situation demands immediate attention by the concerned authorities, especially for the small rivers of the world that hold limited river bed resources. The effects of instream sand mining may not be visible immediately after the extraction processes, because the adverse effects of sand mining require a decade or more to surface and propagate the negative externalities along the river channel in measurable levels (Sreebha 2008). In other words, mining may continue for years without apparent effects in upstream or downstream, only to have geomorphic effects manifest later during high flows. The equilibrium of a river channel is disturbed by sand extraction as material load moving within a dynamic system is intercepted, which triggers an initial morphological response to regain the balance between supply and transport of sediment. Most of the documented studies on the geomorphic effects of river sand mining have reported, extraction lasting over decades that results in large measurable changes in channel form (Fig. 8.1; Plate 8.1). The

Plate 8.1 Erosion of sand and channel incision are the direct effects of clandestine river sand mining. Many structures constructed for rural water supplies are adversely affected by this process. Changes in river morphology and channel incision noticed in Pamba (**a**) and, (**b**) and, Periyar (**c**) and (**d**) rivers during the periods 1999–2009 and 1998–2008

nature of channel adjustments is largely dependent on the nature of mining. In-stream mining generally causes channel bed adjustments, whereas floodplain mining results in more pronounced changes in channel position, associated largely with cut-offs (Mossa and Mclean 1997). The mining activity can have a substantial effect on river systems due to alterations in the spatial distribution of energy and force (Graf 1979). Thus, it is of immediate concern to protect rivers and its pristine water resources by following definite guidelines that regulate and control river sand mining activities within the resilience capacity of the water systems.

8.2 Environmental Laws Pertaining to Sand Mining

Sand mining is an activity that supports one of the largest industrial sectors—the construction industry in the world. It is a fact that the growth rate of construction sector is an important indicator of development. The global value of construction is expected to reach US$12 trillion per annum in 2020 or about 13 % of global GDP (Global Construction Perspectives 2012). In India, the construction industry has accounted for around 40 % of the developmental investment during the past 50 years. This makes absolute restriction on sand mining not a viable option as direct effects would be reflected on the growth and development indices of nations. Every nation, thus, have binding legal policy regulations to control and monitor sand mining operations within their jurisdictions pertaining to stipulated limits and conditions. A few examples of sand and gravel extraction controls enacted or followed by a few nations across the globe are described in this chapter.

The mechanisms to address the issues of resource extraction (major and minor) are essentially important in the sustainable growth of nations as demand of resources is directly proportional to the rate of development. In pursuance to this end, the Indian Parliament enacted the Mines and Minerals (Development and Regulation) Act, 1957 for the regulation of mines and development of all minerals in India including the building materials that comes under the minor mineral category. The Mines Act 1952 and, the Mine and Minerals Act 1957 as well as the rules framed under these two acts are the basic legislations steering the mining sector in the country. A National Mineral Policy was announced in 2008 and the proposal for a new mining legislation—Draft Mines and Minerals (Development and Regulation) Bill was framed for the parliament's consideration in December 2011. The environmental and forest conservation laws also influence the resource extraction sector in the country. Extraction of river sand and gravel from the river beds and floodplains has the potential to seriously disrupt many components of riverine systems. At the same time, aggregates are also economically significant resources for the construction sector. Accordingly, the government has not banned sand extraction from rivers, but regulates the activity to minimize adverse effects. Section 15(1) of the MMDR Act empowers the state government to make rules for regulating the mining/quarrying of minor minerals. In an attempt to contain the adversities of uncontrolled river sand mining, the Government of Kerala enacted the legislation "The Kerala Protection of River Banks and Regulation of Removal of Sand Act, 2001". The act also envisaged provisions for periodical sand auditing to be undertaken to evaluate the feasibility of sand mining operations in the rivers of the state. Several other legal bindings like The River Bank Development Plan prepared by the District Level Expert Committee (DLEC) also contributes to establishing, coordinating, and protecting river banks within the district. The River Management Fund (RMF) maintained by the District Collector is meant for meeting all the expenses toward management of the river systems. In addition to the Act of 2001, provisions under the Kerala Land Conservancy Act, 1957 and The Kerala Minor Mineral Concession Rules, 1967 are also effectively used to protect river banks and river beds from large scale mining /dredging of river sand and to protect the biophysical environments of the rivers subjected to sand mining.

Similar to the Indian scenario, Sri Lanka has declared sand to be a property of the state, which requires a permit to mine and transport [Mines and Minerals Act No. 33 (1992) of Sri Lanka]. Artisanal sand mining was the norm until the introduction of the current Mines and Minerals Act (which replaced the former Mines and Minerals Law No. 4 of 1973) established by the Geological Survey and Mines Bureau (GSMB), which regulates the exploration for and mining of minerals, including sand. The GSMB is responsible for identifying the locations and quantities of sand deposits. The Mines and Minerals Act No. 33 of 1992, the National Environmental Act of 1980, the Coast Conservation Act of 1981, and other relevant legislations, regulations, and policy statements reflect Sri Lanka's constitutional, international, and national obligations toward protecting their natural resource base. These policies support other national policies such as the National Environmental Policy (2003).

In the case of Victorian government, approval is required from the Catchment Management Authority (CMA) wherever extractive works will interfere with the bed or banks of a waterway, or within the floodplain inundation zone, where the CMA has floodplain management functions. Approval is also required from Department of Primary Industries (DPI) under the Extractive Industries Development Act 1995, except where the depth of extraction is less than 2 m below the natural surface and total area of extraction is less than 2,000 m^2. Approval from DPI is not necessary where sand and gravel is extracted from the floodplain for use on the property of origin, and is not for sale or any other commercial use. However, when the mined area is greater than 2,000 m^2 approval will be required under the Catchment and Land Protection Act, 1994 from DPI. Generally, extractive works within waterways is only permitted if the activity is a component of a regional waterway management strategy, or if it can be demonstrated to the CMA that there are clear net gains to the environment or stability of the waterway. In view of adverse impacts on the river systems, the in-stream removal of sand and gravel is generally discouraged throughout Victoria by general CMA policy.

The New South Wales government has put forth "The Sand and Gravel Extraction Policy for Non Tidal Rivers" in the year 1992, which is a component of the State Rivers and Estuaries Policy. The policy encompasses a suite of components, each of which will focus on the management of river sand and gravel extraction for commercial purposes. The commitment of the government toward the protection of riverine and estuarine systems is reflected by a number of legislative controls administered by various government agencies toward this end. The Environmental Planning and Assessment Act, administered by Department of Planning in conjunction with local government have very specific requirements for all extractive industries including river sand and gravel. The Rivers and Foreshores Improvement Act require that a permit be obtained prior to any excavation on freehold protected river land by a private person or a company. The protected riverine area includes land within channel, or within 40 m of the banks of any water course. Several other acts like The Crown Lands Act, The Forestry Act, and The Soil Conservation Act work in concurrence to minimize and control the aggregate extraction activities within certain limits.

The Malaysian government has also put forth strict guidelines and recommendations that have to be followed by the river sand extraction industry of the country. Accordingly, sand dredging must be at a minimum distance of 10 m from the main channel bank toward the flow channel. The stockpile must be located beyond 30 m to the left and right of the main channel bank. The minimum depth of the excavation or redline must be 1 m above natural channel thalweg elevation.

The maximum allowable mining depth is 1.5 m. In addition to in-stream mining, the Ministry of Natural Resources and Environment, Department of Irrigation and Drainage Malaysia (2009) have also come forward with a set of conditions for flood plain mining as well. The excavation must be setback at a minimum of 50 m distance from the main channel bank. The maximum depth of floodplain extraction should remain above the channel thalweg. Side slope of the floodplain excavation should range from 3:14 to 10:1. The whole procedure is made for improving the overall quality of the rivers in Malaysia.

Over the last few decades, it has become increasingly obvious that many rivers are being over-extracted for river sand resources and significant channel enlargements have resulted in the global context. Countries that have successfully implemented the sand mining rules and regulations at grass root level alone have recorded successful protection of river systems in the long run. The developing nations of the world will need to strengthen their policy implementation mechanisms to successfully tackle illicit sand mining operations in their riverine zones to rescue the river environments from further degradation.

8.3 Purpose of the Guidelines

The guidelines are intended to be put forth to use by the authorities concerned, and to update the existing sand and gravel permitting policies for achieving the goal of sustainable extractive management practices. The purpose of these guidelines is to ensure how sand mining can be developed to realize its economic and social potential in a sustainable way. Though the guidelines are intended for use principally by policy makers and practitioners, it can also be of immense use to operators of sand mining as well. The guidelines aim to achieve a consistent approach for promoting sustainable sand extraction integrating management of water and land resources in line with the principles of catchment management. The following are some of the major objectives of the guidelines:

- Ensure that the extraction of sand and gravel is undertaken on a sustainable basis,
- Manage extraction processes in a way which minimizes detrimental effects if any, on the river ecology,
- Apply proper sand auditing in maintaining the river equilibrium in tune with sediment replenishment to identify suitable locations, period, and quantity that can be extracted, and
- Ensure that the extraction policy is consistent with the other existing government policies and initiatives.

The entire channel floodplain system is important to fluvial ecosystem functions and also managing river health. These guidelines address floodplain and terrace pits, in addition to instream mining, because such pits may capture the sediment load of adjacent streams, and may affect water quality and quantity in nearby streams.

8.4 A Framework for Action

In recent years, there has been increasing awareness and concern about the importance of managing natural resources of rivers and their floodplains, collectively known as alluvial sources of sand. The main frame of action is to ensure that river sources continue to support the socio-economic and environmental uses of sand. The most urgent necessity in controlling indiscriminate sand extraction is to ensure effective, efficient, and purposive administration of the existing sand mining laws. The field-level monitoring and control should receive special care where administrative reforms are being made at grass root level. The local self governments should be made fully responsible and accountable for the sand mining activities in their jurisdiction.

A sustainable development framework along with a set of sustainable indicators is required to enhance public accountability of sand mining operators. An effective framework requires an integrated approach to: developing position statements, objectives and policies, processes, delivery, and actions, and a basis for appraisal. Its key elements could be: a) controlled and scientific mining, b) environment protection and mitigation, c) community stakeholder engagement, and d) transparency and accountability. The government should primarily be concerned with the "legal framework" for sustainable mineral development and ensure that the relevant laws are implemented fairly and effectively in order to ensure good governance in the sand mining sector.

8.5 The Importance of Channel Maintenance Processes

To a large extent, channel maintenance processes govern the channel morphology and the resulting fish habitat, which is altered by river sand mining. Changes to channel maintenance processes resulting from sediment removal can adversely hit proper functions of river system. In the interest of protecting the rivers, it is undesirable for channel disturbances to widely alter channel conditions from the range of the channel-forming (effective) flows. The effective discharge is the flow that is most effective in the long-term transport of sediments (Wolman and Miller 1960), and have been used to determine the equilibrium status of channels (Florsheim et al. 1998), to quantify channel maintenance flows (Nash 1994), and also to specify instream flow requirements (Schmidt and Potyondy 2001; Andrews and Nankervis 1995). Maintaining equilibrium channel size and sinuosity requires that the sediment transport capacity of the channel is, on average, matched to the supply from uplands, so that in the long run, the channel will set to become stable—ie, neither degrades nor aggrades (Emmett 1999). This assumes an available supply of sediment and if there is not adequate supply, then transport causes incision. Therefore, channel-forming processes are most effectively conducted by the flow that transports most sediment load over time (Wolman and Miller 1960; Leopold et al. 1964; Knighton 1984). In a systematic study on sand mining activities of central Kerala, SW India, Sreebha (2008) reported the rivers in their storage zones exhibit marked “human imposed” modifications consequent to paucity of sediment supply from upstream and midstream reaches. There is therefore an urgent need for river restoration to revert the already disturbed natural equilibrium in terms of sediment supply from the river catchments (Maya et al. 2012).

8.6 Management Options and Extraction Control Measures

The most effective way to protect, or restore health and stability of rivers are by protecting naturally occurring physical processes that create and maintain the river systems. Rivers can be protected from the negative impacts of sand mining by implementing a combination of two methods that minimize the disturbance of stream channel: a) minimizing channel modifications and b) limiting the volume of sand extraction below the level of natural replenishments. It is important that sediment extraction operates at scales that do not intercept the quantity of coarse sediment supply. Several researchers have reported different methods to control sediment extraction from rivers. Kondolf (1994) opined that aggregate extraction from rivers must be managed on a river basin-scale, while the demand for aggregate and its supply must be managed on the production-consumption basis. According to Kondolf et al. (2002), one strategy is to define a “redline”, a minimum elevation for the thalweg and to permit mining so long as the bed does not incise below this level. This can be verified by yearly or half yearly river profile surveys by expert groups. The channel bars along some reaches of large rivers constitute a renewable resource of sand and gravel, which can be extracted without significant damage to the channel if proper care is taken while permitting the sand extraction processes. Methods developed based on “Redline” concept is used as a base in many countries to regulate the sand extraction process. When mining touches the red line, in subsequent years only the aggregate that is replenished above the “redline” is allowed to be extracted. This concept is applicable in river reaches where sand aggradation is noticed. During monsoon (high flow periods), deposition above redline can be voluminous, while during summer there may not be any significant deposition, especially the case of rivers in the non-glaciated regions. One advantage of the management method based on the “redline” concept is that it can allow for varying climatic and sediment transport events.

Sediment budget methods are used in some areas to limit the volume of material involved in commercial sediment removal operations. Regulating the extraction by sediment budget methods allows fairly consistent annual extraction rates even though sediment delivery depends on decade to century level cycles. However, regulating extraction to sediment budget does not provide for maintenance of geomorphic features that serve ecological functions including fish habitat in a river (Paukert et al. 2008; Rempel and Church 2009). Poorly planned and careless extraction of sand from river channel bars can have deleterious effects on bar stability, channel evolution, and various other aspects of river characteristics. Fish habitat, aesthetics, and other river characteristics may be detrimentally affected as an indirect result of the channel disturbance. Periodic field checking of river sand mining sites, quantity of sand extraction, periodic environmental auditing, etc., could prevent degradation of river channels.

Current approaches for managing instream mining are based on empirical studies. Sediment transport models can provide an indication of potential channel incision and aggradation, but all such models are simplifications of a complex reality, and the utility of existing models is limited by unreliable formulation of sediment rating curves, variations in hydraulic roughness, and inadequate understanding of the mechanics of bed coarsening and bank erosion (NRC 1983). Another approach to manage sand mining is to estimate the annual bedload sediment supply from upstream, in various sections of the storage zones and to limit sand extraction to that value or some fraction of it (Kondolf et al. 2002).

8.7 Restoration and River Management

Restoration of rivers can be achieved by bringing down the level of sand extraction and allowing the system to recuperate by itself. Restoration is the re-establishment of the structure, functions, and natural diversity of an area that has been altered from its natural state by human interferences (Cairns 1988; NRC 1992). Restoration projects are beneficial to the society, but many well-intentioned projects are ineffective or detrimental due to lack of proper planning and implementation (e.g., Frissell and Nawa 1992; Iversen et al. 1993; Kondolf et al. 1996). A competent geomorphological analysis can shed light into the fluvial processes and controls at a catchment scale, and a historical analysis can document the evolution of the channel and catchment, providing the manager with insight into the underlying causes of the channel's current condition (Sear 1994). River geomorphology and ecology are complex, and one cannot predict precisely how the river will respond to a given set of treatments. From an environmental point of view the best restoration method is the one, which is absolutely necessary and involves minimum modification of the natural channel.

8.8 Management Plan

Removal of a stream's bedload disrupts the sediment mass balance and alters channel geometry and elevation. From geomorphic principles, it can be predicted that sediment removal should induce predictable channel responses and corresponding changes to riverine habitats. These guidelines provide recommendations for the evaluation, design, and monitoring of sediment removal activities in

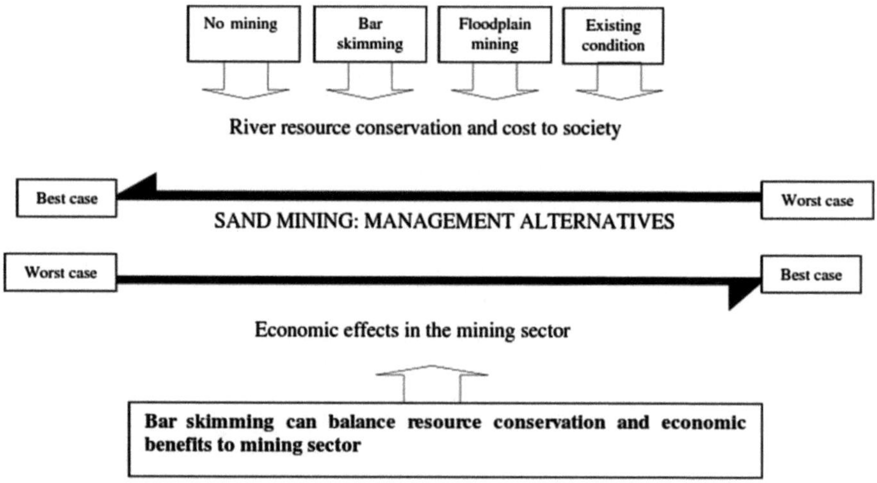

Fig. 8.2 Management alternatives of sand mining (Sreebha and Padmalal 2011)

streams. Many countries have implemented strict regulations, including prohibition of instream sand mining and rigorous planning and monitoring measures as requirements for obtaining permits. Despite these measures, illegal mining continues to plague the mining sector. The following policies are to be considered before issuing sand and gravel mining permits:

- Ensure conservation of the river equilibrium and its natural environment.
- Ensure that the rivers are protected from bank and bed erosion beyond its stable profile.
- Avoid interference on the river maintenance work or other structures like bridges.
- Avoid obstruction to the river flow, water transport, and river water problem.

A management plan is prepared to mitigate the adverse impacts of sand mining (see Fig. 8.2). Instream mining can be managed by adopting any one of the four alternatives mentioned in the schematic model: (1) no mining in channels or floodplains, (2) bar skimming only, (3) floodplain mining only, and (4) no change to existing regulations. These alternatives range from best case (1) to worst case (4) in terms of river resource conservation and costs to society and from worst case to best case for economic effects on the industry.

The management plan outlines environmental protection and other measures that will be undertaken to ensure compliance with environmental laws and regulations and to reduce or eliminate adverse impacts. A share of the revenue collected from sand mining should be used for river restoration since the well being of society is essentially a function of the environment.

8.8.1 Instream Mining

Continuous monitoring is essential in measuring actual replenishment rate of rivers, so that sand extraction is well within the rates of replenishment, which protects long-term channel stability as well as aquatic and riparian habitat by extracting a volume sustainable by watershed processes. As mentioned, the redline concept put forth by Kondolf et al. (2002) can be used as absolute elevation below which no mining could occur. Surveys need to be site-specific in order to avoid impacts on structures and to avoid vegetation impacts associated with down cutting by excessive removal of sediment. An extraction site can be determined after setting the deposition level at 1 m above natural channel thalweg elevation, or as determined by the approved agency.

Of the different alternatives, bar skimming (extraction of sand from the exposed sand deposits/bars within river channels) can, to some extent, balance resource conservation and economic benefits to mining sector. Controlled bar skimming may be considered in the river reaches where sand bars are seen exposed and well developed. Even though bar skimming is recommended as a means for advancing river conservation and maintaining a viable extraction industry, no mining should be opted for already sand depleted rivers for a certain period of time in order to restore them back to their natural condition. Other methods such as excavation of trenches or pools in the downstream channel lower the local base level.

Retaining the upstream one to two thirds of the bar and riparian vegetation while excavating from the downstream third of the bar is also accepted as a method to promote channel stability and protect the narrow width of the low flow channel necessary for habitat preservation. In contrast, if excavation occurs on the entire bar after removing existing riparian vegetation, there is a greater potential for widening and braiding of the low flow channel. In-stream extraction activities should be concentrated or localized to a few bars rather than spread out over many bars. No washing, crushing, screening, stockpiling, or plant operations should occur at or below the streams "average high water elevation," or the dominant discharge (DID, 2009). The cumulative impact of all dredging activities should be reviewed on an annual basis to determine if riverine effects are likely to be well within the resilience of the system. Ensure that permits are distributed in a manner that minimizes long-term impacts and inequities in permits between adjacent mining operations.

8.8.2 Floodplain Mining

In areas where sand and gravel occurs on floodplains or terraces, there is a potential for the river channel to migrate toward the pit. If the river erodes through the area left between the excavated pit and the river, there is a chance for "river capture", a situation where the low flow channel is diverted though the pit. In order to avoid river capture, excavation pits should be set back from the river to provide enough buffer area. The depth of the excavation must be above natural channel thalweg elevation, as determined by the concerned authorities and based on appropriate databases. Floodplain pits should not be excavated below the elevation of the thalweg in the adjacent channel. This will minimize the impacts of potential river capture by limiting the potential for head cutting. A shallow excavation (above the water table) would result in the formation of a pit that would fill with water part of the year (DID, 2009).

The excavated area must allow for a 3:1 slope from the buffer zone to the bottom of the excavation. For the streamlined permit, it is not acceptable to excavate vertically to the buffer zone and propose to backfill the excavation to achieve the required 3:1 slope. Terrace pits should be designed with a large percentage of edge habitats with a low gradient, which will naturally sustain vegetation at different of water levels. A reclamation plan is required for streamlined permit applications in the case of flood plain mining.

8.9 General Guidelines

a. In the case of mining leases for riverbed sand mining, specific river stretches should be identified and mining permits/lease should be granted stretch-wise. Safeguard measures should be duly implemented and effectively monitored by the regulatory authorities.
b. For mining in the proximity of any bridge and/or embankment, appropriate safety zone should be worked out on case to case basis, by taking into account the river characteristics and the environmental effects of sand mining.
c. Proper inventory of the available sediment resources through sand auditing should be made mandatory prior to sanction of leases (an example from Kerala state in the southwestern coast of India is given in Chap. 9).
d. River reaches that experience deposition or aggradation shall be identified first. Sand and gravel extraction may be allowed in such locations to lessen aggradation problems in the overbank.
e. A pre-dredging baseline survey and dredging and post-dredging monitoring should be incorporated in the policy guidelines.
f. The distance between sites for sand and gravel mining shall depend on the replenishment rate and width of the river.
g. Sand and gravel shall not be allowed to be extracted where erosion may occur, such as in the concave banks.
h. Dredging equipments that minimize siltation threats and turbidity levels in the overlying water should be encouraged.
i. Dredging is to be conducted during periods of lowest biological activity, and periodic biota surveys during and after the dredging operation must be conducted.

j. To the possible extent, clearing and removal of vegetation and timber from the riparian zone should be avoided in preparing the mining site and access routes.
k. An undisturbed vegetated buffer of at least 100-feet in width landward from the top of the riverbank shall be maintained, and mining must be operated in such a manner so as to allow no discharges to the river.
l. No dewatering piles or stockpiles shall be positioned within the banks of the channel below the Ordinary High Water Mark.
m. To the possible extent, the operator must implement a "one-step removal" process. This normally involves excavating or dredging the material with a bucket-type loader. Construction, of approach roads interrupting the flow or wetted portion of the waterway channel should be strictly avoided.
n. The extraction permit shall be attentive to seasonal flow conditions and minimize removal strictly pertaining to legal framework.
o. The authorities should also be aware of local spawning seasons and conditions, and restrict operations within the waterway channel during those times.
p. Planting trees along the riverbanks with no or minimal vegetation, irrespective of signs of erosion (indigenous species) should be advocated as a compensatory measure. Preferably, the natural vegetation and land use practices that existed before mining shall be reintroduced.

References

Andrews ED, Nankervis JM (1995) Effective discharge and the design of channel maintenance flows for gravel-bed rivers. In: Natural and anthropogenic influence in fluvial geomorphology, geophysical monograph, vol 89. American Geophysical Union, Washington DC, pp 151–164

Cairns J Jr (1988) Rehabilitating damaged ecosystems. CRC Press, Boca Raton, p 405

DID (2009) River sand mining and management guideline. Department of Irrigation, Malaysia, p 50

Emmett WW (1999) Quantification of channel-maintenance flows for gravel-bed rivers. In: Wildland Hydrology, vol 99. American Water Resources Association, Herndon, pp 77–84

Florsheim J, Goodwin P, Marcus L (1998) Geomorphic effects of gravel extraction in the Russian River, California. In: Bobrowsky PT (ed) Aggregate resources: a global perspective. Balkema Publishers, Rotterdam, Netherlands, pp 87–99

Frissell CA, Nawa RK (1992) Incidence and causes of physical failures of artificial habitat structures in streams of western Oregon and Washington. North Am J Fish Manag 12:182–197

Global Construction Perspectives (2012) www.globalconstruction2020.com. Accessed 13 Dec 2012

Graf WL (1979) Catastrophe by theory as a model for change fluvial systems. In: Rhodes DD, Williams GP (eds) Adjustments in the fluvial system. Dubuque Kendall/Hunt Publishers, Dubugue, Iowa, pp 13–32

Iversen TM, Kronvang B, Madsen BL, Markmann P, Nielsen MB (1993) Re-establishment of Danish streams: restoration and maintenance measures. Aquat Conserv Mar Freshw Ecosystems 3:73–92

Knighton D (1984) Fluvial forms and processes. Edward Arnold/Hodder and Stoughton, London, p 320

Kondolf GM (1994) Environmental planning in the regulation and management of instream gravel mining in California. Landscape Urban Plann 29:185–199
Kondolf GM, Smeltzer M, Kimball L (2002) Freshwater gravel mining and dredging issues. Report prepared for Washington department of fish and wildlife, Washington department of ecology and Washington department of transportation, 122
Kondolf GM, Vick JC, Ramirez TM (1996) Salmon spawning habitat rehabilitation on the Merced river, California: an evaluation of project planning and performance. Trans Am Fish Soc 125:899–912
Leopold LB, Wolman MG, Miller JP (1964) Fluvial processes in geomorphology. WH Freeman, San Francisco, p 522
Maya K, Padmalal D, Babu KN (2012) Environmental degradation of Muvattupuzha river basin-causes, consequences and strategies for river restoration. Technical Report, Centre for Earth Science Studies, Thiruvanathapuram, Kerala, p 110
Mossa J, McLean M (1997) Channel planform and land cover changes on a mined river floodplain—Amite River, Louisiana, USA. Appl Geogr 17:43–54
Nash DB (1994) Effective sediment-transporting discharge from magnitude-frequency analysis. J Geol 102:79–95
National Environmental Policy (2003) Draft National Environmental Policy request for public views and comments. Ministry of Environment and Natural Resources of Sri Lanka, p 13
National Research Council (NRC) (1983) An evaluation of flood-level prediction using alluvial river models. Committee on Hydrodynamic Computer Models for Flood Insurance Studies, Advisory Board on the Built Environment, Commission on Engineering and Technical Systems, National Academy Press, Washington DC
National Research Council (NRC) (1992) Restoration of aquatic ecosystems: science, technology and public policy. National Research Council, National Academy Press, Washington DC, p 62
Paukert C, Schloesser J, Fischer J, Eitzmann J (2008) Effect of instream sand dredging on fish communities in the Kansas river, USA: current and historical perspectives. J Freshw Ecology 23:623–634
Ramachandran KK, Padmalal D (1997) Environmental Impact Assessment (EIA) of mineral based industries. In: Nair, Nair (eds) Natural Resources of Kerala, Balachandran Thampi, WWF for Nature, Thiruvananthapuram, pp 31–49
Rampel LL, Church M (2009) Physical and ecological response to disturbance by gravel mining in a large alluvial river. Can J Fish Aquat Sci 66:52–71
Schmidt LJ, Potyondy JP (2001) Flow regimes needed to maintain channels in gravel-bed rivers. In: Proceedings of the 7th Federal interagency sedimentation conference, Reno, Nevada, vol 1, pp 66–69
Sear DA (1994) River restoration and geomorphology. Aquat Conserv Mar Freshw Ecosystems 4:169–177
Sreebha S (2008) Environmental impact of sand mining: a case study in the river catchments of Vembanad lake, Southwest India. PhD Thesis (unpublished), Cochin University of Science and Technology, Kochi
Sreebha S, Padmalal D (2011) Environmental impact assessment of sand mining from the small catchment river in the southwestern coast of India. Environ Manage 47:130–140
Wolman MG, Miller JP (1960) Magnitude and frequency of forces in geomorphic processes. J Geol 68:54–74

Chapter 9
River Sand Auditing: An Example from SW India

Abstract Mining of sand and gravel with little regard to natural replenishment processes have imposed marked changes on the biophysical environment of the small rivers of Kerala in the southwest India. Considering the importance of management of river bed resources, the Government of Kerala enacted the legislation "The Kerala Protection of Riverbanks and Removal of Sand Act, 2001" to protect the rivers from large scale sand mining. This act also envisages provisions for periodic sand auditing in order to evaluate the sand mining process and to protect the rivers from unscientific human interventions. This chapter deals with the various procedures employed in sand auditing of the small rivers of Kerala (southwest India), taking the case of Periyar river an example.

Keywords River sand auditing · Mineable sand estimation · Resource allocation · Performance evaluation of sand mining

9.1 Introduction

Rivers and the adjoining wetland ecosystems of the world are fast degrading consequent to indiscriminate sand and gravel mining from river beds and related areas. Evidences of the adverse environmental effects of river sand and gravel mining are increasing year after year as researchers across the world are focusing more on studies related to human interferences in river ecosystems. Crisis of potable water availability makes the problems in rural water supply schemes attached to rivers more pressing than ever before. Developed nations spend millions of dollars for regenerating river systems, damaged by sand and gravel extraction. The situation of rivers is now critical in many developing nations of the world. The problem is rather worse in regions that are drained by small rivers, as the sand and gravel resources in such rivers are very limited. One of the best examples is that of Kerala state in southwestern India. The state is fast developing owing to the upsurge of economic growth, wise utilization of human resources and

D. Padmalal and K. Maya, *Sand Mining*, Environmental Science and Engineering,
DOI: 10.1007/978-94-017-9144-1_9,

liberalized banking policies. All these, in one way or the other, have accelerated the pace of building constructions and other infrastructural developments in the state. This has led to large-scale extraction of building materials from the state's limited natural resource base. Indiscriminate sand and gravel extraction from the river environments beyond the natural resilience has imposed irreparable damages to all its 41 west flowing rivers. Considering the importance of managing the natural resource extraction from rivers and their floodplains, the Government of Kerala enacted the legislation "The Kerala Protection of River Banks and Regulation of Removal of Sand Act, 2001" to protect river banks and river beds from large scale dredging of sand. This act envisages provisions for periodic sand auditing in order to evaluate the sand mining processes and to protect the rivers in the state from further degradation.

9.2 River Sand Auditing

Sand auditing is a procedure to evaluate the process of sand mining in a river or a portion of river after a specific period of sand mining with an aim to lay down strategies for improving the overall environmental quality of the river. More specifically, this exercise has to be carried out to know how the mining processes and its execution would minimize the negative effects of sand mining on one hand and maximize positive effects on the other. The audit report has to provide necessary recommendations and guidelines regarding the future of sand mining in the river or a part of the river which is subjected to sand auditing.

9.3 Methodology

The sand audit methodology developed and adapted by Padmalal et al. (2010) for the Manimala river draining into the Vembanad lagoon in the southwest coast of India has three major components.

Component I: Resource estimation
Component II: Resource allocation
Component III: Performance evaluation of sand mining.

9.3.1 *Resource Estimation*

A realistic assessment of sand resource is essential for estimating the mineable limits of sand in each stretch of river environment. The following steps are suggested for the purpose:

1. Mapping the river channel and associated natural (instream and riparian vegetations, sand bars, pools, riffles, etc.) and artificial structures/features (bridges, water intake structures, side protection structures, etc.).
2. Estimation of sand resource in the river channel using suitable methods (shallow seismic surveys in river stretches with sufficient depth and width to run the instrument, resistivity surveys in dry river beds, spiking and coring using specially fabricated coring devices, etc.).
3. Estimation of mineable quantity of sand in the river channel. "Mineable quantity of sand" is the volume of sand resource up to a specific depth (fixed by an expert group) in river reaches other than the following areas:

- Prohibited areas mentioned in "The Kerala Protection of River Banks and Regulation of Removal of Sand Act, 2001."
- River reaches with channel width less than 50 m and water depth greater than 3 m,
- River reaches identified as biological hotspots/biologically significant reaches confirmed by experts,
- River reaches with alluvial placers or any other economic minerals,
- River reaches critically affected by bank sliding/slumping, caving, bank erosion, etc.,
- Any other feature(s) of the river environment that the expert team decides to be protected in the river, which is subjected to sand auditing.

9.3.2 Resource Allocation

The steps to be adopted for this purpose are described below:

Let 'X' be the total quantity of mineable sand in million cubic meters (Mm^3), 'Y' the annual replenishment of river sand in Mm^3, and 'N' the time span in years during which mining of river sand can be permitted in the entire river or that part of the river under examination. Then,

Quantity of mineable sand in a year (Q_m) can be estimated as

$$Q_m = (X/N) + Y \tag{9.1}$$

Note 'N' should be fixed only after taking into account the physical, chemical and biological status of the river environment by the expert group. 'Y' can be calculated by subtracting quantity of sand output (Q_o) from the quantity of sand input (Q_i), (see Fig. 9.1)

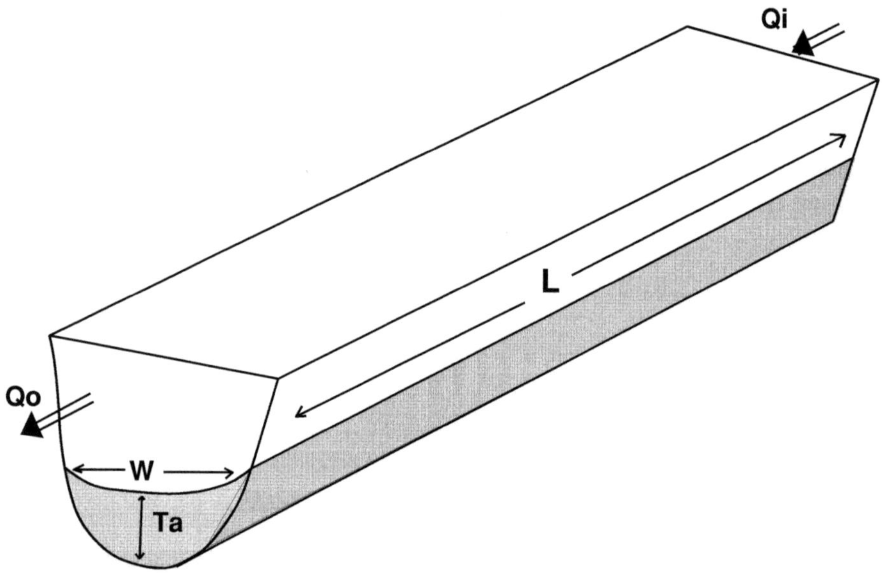

Fig. 9.1 Schematic model of channel segment in the storage zone of a river. *Note* Q_i and Q_o will be determined either using modeling studies or by direct measurements using bedload samplers/traps and suspended load samplers. T_a can be determined from monthly/seasonal cross-profile measurements of river channel. Q_i quantity of sand input into the river/river segment under examination, Q_o quantity of sand output from the river/river segment, L length of the channel occupied by sand, W width of the channel occupied by sand, T_a actual river bed lowering

9.3.3 Performance Evaluation/Sand Resource Accounting

Sand resource accounting is a scientific evaluation of the performance of sand mining activities during the period of sand auditing in a given stretch of river channel or the entire river system.

Preparation of an assessment of mineable sand resource after a specific period of sand mining requires the following input data, if the river maintains its continuum.

Assume that sand mining is uniform throughout the river stretch or riverbed is being leveled after every peak flow season (monsoon),

Then,

Quantity of sand mining

$$Q_m = (Q_i - Q_o) + Q_c \tag{9.2}$$

where, Q_c is the quantity of sand mined from the deposit other than natural replenishment in the given stretch of river under investigation. From Eq. (9.2), Q_c can be calculated as

$$Q_c = Q_m - (Q_i - Q_o) \tag{9.3}$$

Mining of Q_c from the rivers will be reflected in the seasonal cross-profile measurements as channel incision/riverbed lowering. The expected riverbed lowering (T_e) due to mining of Q_c from the river segment under examination can be calculated as

$$T_e = Q_c/LW \tag{9.4}$$

Apply the value of Q_c in Eq. (9.4). Then,

$$T_e = [Q_m - (Q_i - Q_o)]/LW \tag{9.5}$$

Ideally, the computed T_e will be equal to the actual riverbed lowering (T_a) obtained from river cross-profile measurements, provided there is no unauthorized mining in the river stretch.

In other words, a situation in which $T_a > T_e$ indicates prevalence of unauthorized mining in the stretch, the quantity of which (i.e., Q_{um}) can be calculated as:

$$Q_{\mathrm{um}} = LWT_a - Q_m \tag{9.6}$$

Similar survey and computational procedures of river sand may be extended to the entire alluvial reaches of river systems or particular river segments for sand auditing.

9.3.4 Estimation of Unauthorized Sand Mining: An Example

This section deals with estimation of unauthorized sand mining from Periyar river in the southwest coast of India. Estimation of unauthorized sand mining in a given stretch of a river is a difficult task because of the non-availability of continuous yearly data of channel profile measurements, discharge through the river segment, information on the quantity of authorized sand mining, etc. As a fairly good database of these parameters was available for the Periyar river between Malayattoor, and Neeleswaram, an attempt has been made in this segment to compute the extent of unauthorized sand mining during the period 2001–2005.

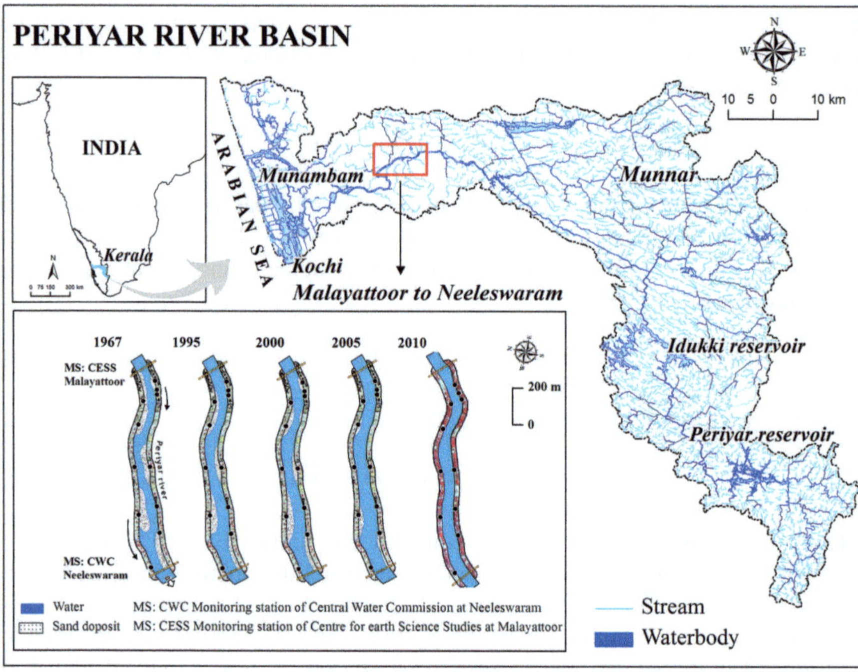

Fig. 9.2 The channel segment between Malayattoor and Neeleswaram showing vanishing of sand deposits over the years

The length of the river stretch (*L*)	= 4,536 m
Average width of the river channel (*W*)	= 267 m
Average channel incision/river bed lowering (T_a)	= 0.82 m
Quantity of sand input (Q_1) computed at CESS monitoring station at Malayattoor based on Meyer-Peter's equation (*Data source* Binoy 2011)	= 66,660 m^3
Quantity of sand output (Q_0) computed in CWC station (*Data source* Binoy 2011)	= 63,920 m^3
Quantity of authorized sand mining from the stretch as per available records	= 648,800 m^3

Expected river bed lowering computed for an amount of 64,880 m^3 as per Eq. (9.5), for the period 2001–2005

$$
\begin{aligned}
T_e &= [Q_m - (Q_i - Q_o)]/LW \\
&= 648{,}800 - (66{,}660 - 63{,}920)/4{,}536 \times 267 \\
&= 0.533\,\mathrm{m}
\end{aligned}
$$

But the actual riverbed lowering estimated from channel surveys for the stretch, Neeleswaram-Malayattoor (T_a) = 0.82 m. This means that the value T_a in this case is greater than T_e (0.533 m). This is an indication of unauthorized sand mining that has taken place in the river stretch during the period 2001–2005.

As per Eq. (9.6), the quantity of unauthorized sand mining is

$$\begin{aligned} Q_{\text{um}} &= LWT_a - Q_m \\ \text{i.e., } Q_{\text{um}} &= (4{,}536 \times 267 \times 0.82) - 648{,}800 \\ &= 993111.84 - 648{,}800 \\ &= 344311.84\,\text{m}^3 \end{aligned}$$

From Fig. 9.2, it is seen that the most notable changes in the exposed sand deposit occurred during the period 2000–2010 and no exposed bars or point bars were noticed in the 2010 scenario.

References

Binoy AM (2011) River sand inflow assessment and optimal sand mining policy development. In: Proceedings of the national seminar on mining of river sand and its impact on the environment, 18–19th Feb, CWRDM, Kozhikode, pp 87–110

Kerala Gazette (2002) The Kerala Protection of River Banks and Regulation of Removal of Sand Act (2001). Kerala Gazette, vol 47, No 285, p 14, Government of Kerala

Padmalal D, Maya K, Narendrababu KN, Baijulal B (2010) Environmental appraisal and sand auditing of Manimala river, Kerala, India. Project Completion Report submitted to Revenue Department, Government of Kerala

Chapter 10
Sand: The Fine Aggregate

Abstract Aggregate materials like sand, gravel, and crushed stone occupy a substantial volume of concrete used in constructions. The demand for high quality aggregates that meet the specifications of civil engineers for construction activities is increasing year after year. A better understanding of the aggregate properties is essential for ensuring the construction processes economical and civil structures long lasting. This chapter deals with a brief description of the characteristics and usefulness of fine aggregates for different construction purposes. The precautions that are to be taken during transportation and storage of aggregates are also discussed in the chapter.

Keywords Fine aggregates · Fine aggregate utility · Properties of aggregates · Storing and handling of aggregates

10.1 Introduction

Natural aggregates are the most valuable non-fuel mineral commodity (Lüttig 1994). It is estimated that about 15 billion tons of aggregate materials worth 76 billion dollars are being produced annually throughout the world (Regueiro et al. 2002). Readily available supply of quality aggregate resources are essential to maintain the infrastructure development of the world. The sterilization of sand and gravel resources due to conflicting land uses is recognized to be one of the greatest challenges in aggregate availability (USGS 2002). The rapid expansion of cities, incompatible land uses, negative impacts of traffic, unacceptable changes on the landscape, and undesirable environmental impacts of developmental activities have contributed immensely to the sterilization of aggregate resources in construction sector. The depletion of aggregates (both on land and river sources) due to encroachment of residential areas and increasing stringency of specifications has

D. Padmalal and K. Maya, *Sand Mining*, Environmental Science and Engineering,
DOI: 10.1007/978-94-017-9144-1_10,

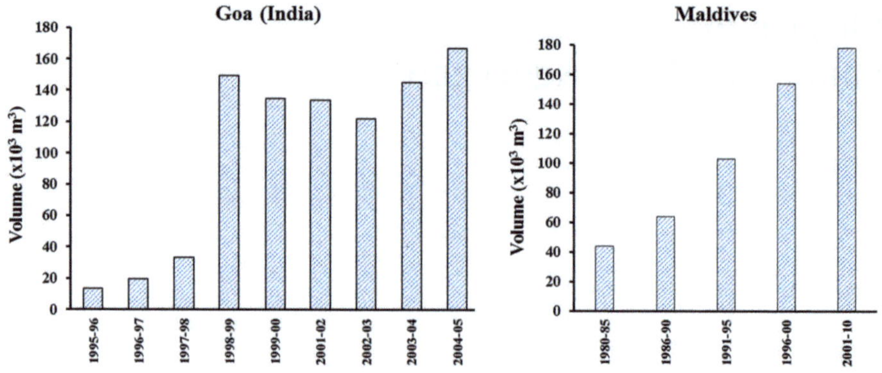

Fig. 10.1 The demand of construction grade sand in Goa, India (after Sonak et al. 2006) and Maldives (after Brown 1997)

necessitated more adequate methods of their exploration and extraction. The problem of aggregates availability has worsened, requiring immediate attention to contain the difficulties that arise from the diminishing resource availability of construction grade sand and gravel.

Aggregates materials such as sand, gravel, or crushed stone usually occupies approximately 60–75 % of the volume of concrete used for various construction purposes. Aggregate properties significantly affect the workability, durability, strength, thermal properties, and density of hardened concrete in civil constructions. The demand for fine and coarse aggregates has escalated rapidly in consonance with the expansion of transportation and construction infrastructure in the mid-twentieth century (Fig. 10.1). Unlike metals, that have high "unit value," aggregate materials are high-bulk, low unit value commodities that derives much of its value from being located near the central market (Bates 1969). In the future, the construction of infrastructure like roads, highways, bridges, airports, seaports; environmental applications such as water and drainage systems; for making steel, glass, and other consumer products; and numerous agricultural, metallurgical, and pharmaceutical purposes will require enormous quantities of aggregate materials (Langer and Glanzman 1993). The demand for aggregate materials is expected to rise exponentially in consonance with population rise and economic development.

10.2 Sand: The Fine Aggregate

Natural sand is a weathered and worn out particle composite derived originally from primary (igneous) and secondary (metamorphic and sedimentary) rocks. The sand that is to be used in construction purposes must have proper gradation of particles from 150 microns to 4.75 mm (Plate 10.1). When fine particles are in proportion, the sand will have fewer voids and the cement quantity required will be

Plate 10.1 River sand and gravel. **a** Poorly sorted river sand with angular to sub-angular grains; **b** Pebbles left as lag concentrate in a point bar deposit

less. Demand for manufactured fine aggregates for making concrete is increasing over the years as river sand cannot meet the rising demand of the construction sector. Because of the limited supply, the cost of natural sand is rapidly increasing. Further, its consistent supply to the construction sector cannot be assured as natural sands are integral components of aquatic ecosystems in nature. Further, poorly graded natural sand with excessive silt and organic impurities is detrimental to the durability of steel in concrete. Under these circumstances, use of manufactured sand becomes inevitable to maintain the pace of developmental initiatives.

Transporting aggregates for long distances can add significantly to the overall price of the product (Leighton 1991). Therefore, aggregate operations are commonly centered near populated areas and /or central markets. Even though natural aggregate sources are widely distributed throughout the world, there are regions where natural sand occurrences are non-existent or are not necessarily available for immediate use (Langer 1988). Some areas are totally devoid of river sources of sand and gravel, and potential sources of crushed stone may occur at depths that make their extraction uneconomical. In other areas, even if sources of natural aggregate are present, it may not meet quality parameters set for specific uses or may react adversely when used in applications such as concrete or asphalt (Langer and Knepper 1998). The quality parameters are determined by the final application, and can restrict the development of otherwise available aggregate materials.

10.3 Usefulness of Aggregates

Aggregates are essential to build and/or maintain infrastructural facilities that are intended for the overall development of a region. Developed countries need to sustain their level of productivity; whereas developing countries have to expand the pace of their development. All these objectives cannot be fulfilled without the continuous supply of aggregates to the core developmental centers and/or central market. The aggregate materials are dredged, mined, or quarried and are used either in their natural state or after proper processing (crushing, washing, and sizing). The utility potential of aggregate resources depends on various geologic and environmental factors that play a crucial role in alterations induced in rock masses including geochemical processes, chemical decomposition, and mechanical disintegration in different climatic zones.

The particles constituting aggregate materials have certain physical and chemical properties that make them acceptable or unacceptable for specific uses. Only aggregates that meet stringent quality specifications such as those for road pavement and concrete may be regarded as good quality. Some of the important aspects of aggregates in concreting are: (a) strength and durability, (b) free of organic impurities, (c) low alkali reactivity with cement, and (d) proper gradation (for good workability and packing of voids).

Suitability of aggregates for specific end-uses can be determined only by laboratory tests to see if specific quality parameters are met; nonetheless, there is an approximate correlation with rock porosity. In general terms, if the water absorption value is less than 2 %, the material is likely to be good quality aggregate, whereas materials with values exceeding 4 % may not be suitable for applications like road pavement and concrete.

10.4 Aggregates for Constructions

Different types of aggregates such as sand, gravel, and crushed stone are commonly combined with binding media to form concrete, mortar, and asphalt. The importance of using the right type and quality of aggregates in concrete cannot be over emphasized. The fine and coarse aggregates generally occupy 60–70 % of the concrete volume and influence markedly the concrete's properties, mixture proportions, and economy. Properties of fine aggregate (sand) affect concrete properties such as durability, strength, thermal properties, unit weight, and surface friction. River sand deposits consist of gravel and sand that can be readily used in concrete with minimal or no processing. Aggregates are usually washed and graded at the processing units to meet the required specifications in type, quality, cleanliness, grading, and moisture content.

Aggregate particles that are friable or capable of being split are undesirable for concreting purposes. Aggregates containing any appreciable amount of flaky, soft, and porous materials are unsuitable for engineering constructions as their low resistance to weathering can cause surface defects such as cracks and breaks. Fine aggregate content is usually 35–45 % by mass or volume of the total aggregate content.

10.5 Fine Aggregate for Making Concrete

Grading refers to the distribution of particle sizes present in an aggregate. Coarse and fine aggregates are generally sieved separately. That portion of an aggregate passing the 4.75 mm sieve and predominantly retained on the 75 μm sieve is called "fine aggregate" or "sand," and larger aggregate is called "coarse aggregate". Coarse aggregate may be available in several different size groups, such as 4.7–19 mm, or 19–37.5 mm. Fine aggregates generally consist of natural sand (river, floodplain, or marine) or crushed stone, whereas coarse aggregates consist of one or a combination of gravels or crushed stone.

Sand is further subdivided into:

- Coarse sand: 4.5–2.0 mm.
- Medium sand: 2.0–0.425 mm.
- Fine sand: 0.425–0.075 mm.

Very fine sands are not recommended for structural concrete and are often uneconomical. Very coarse sand show difficulties in surface finishing of concrete and produces harsh, unworkable cement mixtures but provides strength. Fine sand provides more cohesion than coarse sand and hence, less sand will be needed if fine sand is used. In concrete making, coarse aggregates from rock (of irregular size) will require more sand than rounded coarse aggregates such as river gravel. In general, aggregates that do not have a large deficiency or excess of any size and gives a smooth grading curve will produce the most satisfactory results in concreting. In most cases, the concrete mix can be designed to fit the availability of sand and coarse aggregates.

10.5.1 Sand for Mortars and Plasters

Soft sand is ideal for making mortar and plaster for brick works in buildings and other constructions. Sands of angular type impart higher mortar strength, however, they are unsuitable for brick laying works as they lack plasticity or workability. Mortar made of coarse sand does not adhere easily to bricks during bricklaying. If only coarse sand is available (mountain areas), sieving through a suitable sieve to separate the fine portion for bricklaying is recommended. Generally, it is considered that sand used for mortar for brick work should pass through a sieve of 8 meshes per inch (3.2 mm) and the sand for plastering and pointing must pass through a sieve of 12 meshes per inch (2 mm). Another recommendation for sand used for plaster and mortar work is that the percentage of material that passes through 600 micron sieve should be 40–100% for mortars, and 80–100% for plasters.

10.5.2 Sand for Filling

Sand is also used in construction of buildings that include filling underground floors and also filling behind retaining walls, etc. In each situation, it should satisfy the specific requirements of its end use. Sand used for filling underground floors has to reduce the capillary suction by which water will travel from foundation soil to the floor. This will require coarse sand with large voids between the grains. Whereas in the case of sand required for filling behind retaining walls, the only requirement is that it should be free draining and nonexpansive. Most sands are not expansive and get saturated at low water content. Thus, most free draining sands are suitable for general filling purpose.

10.6 Storing and Handling of Aggregates

Aggregates should be stored and handled carefully so that they remain free from contamination and dirt. Aggregates should also be segregated and moisture content should remain approximately constant. Therefore, a clean dry and hard patch of ground should be selected before the aggregate is received at the site, keeping in view the position of the mixture and convenience of handling.

10.7 Aggregate Resource Protection

A core issue surrounding the sustainable development of natural aggregate resources is the conflict between regional needs and local opposition to resource extraction. Dunn (1983) termed the conflict, and the consequences arising from it, the "Dispersed Benefit Riddle". The riddle is that, the benefits of aggregate development are dispersed over very large areas, but the community where extraction occurs suffers most of the adverse consequences of resource development (Dunn 1983). Thus aggregate resources needs to be protected by promoting orderly and environmentally sound development practices and by introducing aggregate resource protection into local comprehensive planning and land use controls, as initiated by many western nations. Since critical shortages of aggregate deposits are faced by most countries, extraction activities should follow a rational plan that avoids wastage and causes least disruption to quality of life and respective ecosystems.

References

Bates RL (1969) Geology of the industrial rocks and minerals. Dover, New York, p 459

Brown BE (1997) Integrated coastal management: South Asia. Department for International Development (DFID), United Kingdom

Dunn JR (1983) Dispersed benefit riddle. In: Ault CR, Woodard GS (eds) Proceedings of the 18th forum on geology of industrial minerals: Indiana geological survey occasional paper 37, pp 1–9

Langer WH (1988) Natural aggregates of the conterminous United States. US Geol Surv Bull 1594:33

Langer WH, Glanzman VM (1993) Natural aggregate—Building America's future. U.S. Geological Survey Circular 1110, p 39

Langer WH, Knepper DH Jr (1998) Geologic characterization of natural aggregate: A field geologist's guide to natural aggregate resaource assessment. In: Bobrowsky PT (ed) Aggregate resources: a global perspective. AA Balkema, Rotterdam, pp 275–293

Leighton MW (1991) Industrial minerals resource identification and evaluation. In: Bush AL, Hayes TS (eds) Industrial minerals of the midcontinent—proceedings of the midcontinent industrial minerals workshop, U.S. Geological Survey Bulletin 2111, pp 9–20

Lüttig GW (1994) Rational management of the geo-environment—a view in favour of "Geobased Planning". In: Lüttig GW (ed.) Aggregates—raw materials' giant: report on the 2nd international aggregate symposium, Erlangen, pp 1–34

Regueiro M, Martins L, Feraud J, Arvidsson S (2002) Aggregate extraction in Europe: the role of the geological surveys. In: Geological Survey of North Rhine-Westphalia (ed) Raw materials planning in Europe—change of conditions! New perspectives?: Proceedings of third European conference on mineral planning, Krefeld, Germany, October 8–10, pp 187–198

Sonak S, Pangam P, Sonak M, Mayekar D (2006) Impact of sand mining in local ecology. In: Sangeetha S (ed.) Multiple dimensions of global environmental change, The Energy Research Institute, New Delhi, p 324

USGS (2002) Mineral commodity summaries 2002: U.S. Geological Survey Mineral Commodity Summaries, p 197

Chapter 11
Sources of Sand and Conservation

Abstract Construction grade sand is in short supply in many parts of the world. This has resulted in growing demand of sand to maintain the pace of infrastructure development and construction industry. Therefore, there is an immediate need to bridge the gap between demand and supply of aggregates by encouraging the use of low cost, easily available alternatives to river sand. Also measures are to be taken to reduce the use of natural sand to the barest minimum by adopting technologies with lowland—no sand content in construction sector. This chapter gives an overview of the different sources of sand and the environmental problems linked to it during their extraction with special reference to Kerala state in southwest India.

Keywords Alluvial sources of sand · Non-alluvial sources of sand · Manufactured sand · Conservation of sand

11.1 Introduction

In nature, sand occurs as river channel and floodplain deposits, fluvio-glacial deposits, aeolian (wind) deposits, lake deposits and nearshore-marine deposits (Jensen and Bateman 1979). Among these sources, sand from rivers is largely used for meeting its demand in construction industry. As a result of over exploration, river sand is in short supply in many parts of the world and the construction industry faces difficulties in achieving targets. This warrants the need for alternatives to river sand to bridge the gap between demand and supply. Also measures are to be taken to minimize the use of sand by adopting technologies with low sand—no sand content in construction sector.

D. Padmalal and K. Maya, *Sand Mining*, Environmental Science and Engineering,
DOI: 10.1007/978-94-017-9144-1_11,

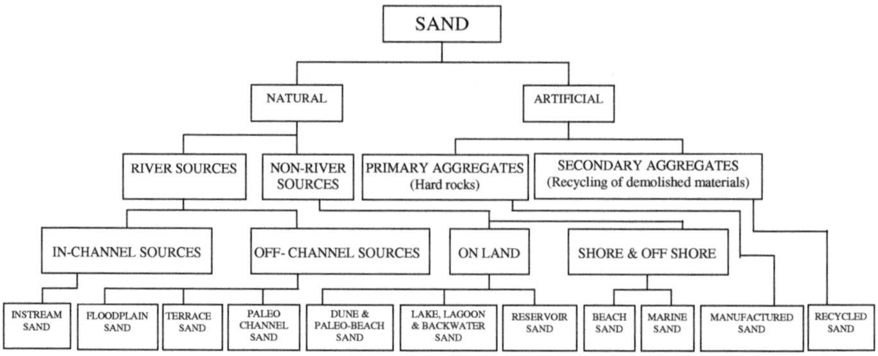

Fig. 11.1 Various sources of sand

11.2 Sand: River Sources

Throughout the developing world, river sand is widely exploited as fine aggregates for building constructions. Sand is either extracted directly from the active river channels (in-channel source/instream source) or from its flood plains or terraces/ overbank areas (off-channel source) (Fig. 11.1). Both these sources are nonrenewable in human life scale. In small rivers of many parts of the world, the in-channel sources of sand are almost depleted and mining of off-channel sources has already imposed severe environmental effects on the socio-environmental scenario of the region, especially in rivers that are close to major developmental centers (Scott et al. 2003; Padmalal 2011). Further, areas where flood plain deposits occur are either the prime agricultural lands or dense settlements. This warrants the need for strict regularization of sand extraction based on the availability of mineable sand.

11.3 Non-river Sources

11.3.1 Dune Sands

The coastal lands of many areas contain sand deposits which are formed initially by littoral processes and later modified into dunes by wind activity. In some cases, palaeo beach ridges (ridges and swales) formed during the regressive phases of sea also occur. Such sands with high content of silica (silica sands) are extracted for glass manufacture or foundry applications. Recently, such sands are also used extensively for building constructions (Padmalal et al. 2004). However, as the dunes near the coast has an ecological and hazard regulatory function its removal has to be prohibited.

11.3.2 Glacial Outwash Sand

Glaciers carry enormous quantities of heterogeneous sediments with them which is kept in a state of transport till the conditions are favorable or till the glaciers reach their terminal points where rate of melting far exceeds rate of movements. The load is then deposited and may form huge accumulations of glacial debris acquiring varying shapes and characters. All such accumulations are called as drifts. Out of the different types of drifts, the stratified drifts contain substantial quantities of sand. Many countries are using sand constituting the fluvio-glacial drift for building constructions after proper processing (Allender and Hollyer 1972). Sorting and grading of sand is required prior to use of such sand.

11.3.3 Sand in Lakes and Backwaters

The coastal lands of many parts of the world are often endowed with a network of fluvial channels, lakes, lagoons, and backwaters. The lakes, backwaters, and lagoons form an inter-phase between rivers and the nearshore areas. These aquatic environments receive sand from both these sources depending on the intensity of river discharge and/or tidal activity. Sand is extracted from the lakes and lagoons in connection with dredging operations. Further, illicit sand mining is also rampant in lakes and lagoons of many parts of the world. This brackish environment has an important ecological function. As an interface between the marine and fresh water systems stringent regulatory controls are warranted for protection of backwater systems from sand extraction.

11.3.4 Beach Sands

Beaches occur in many parts of the coast. Beaches are the natural barriers protecting land areas from coastal erosion. Sand, the granular materials constituting the beach, plays a pivotal role in reducing the fury of the waves in high energy regimes, as it reduces the hungry water effect, i.e., the destructive effect of sand-free waters. It is seen that a substantial quantity of sand from beaches is also mined illegally for various purposes including building constructions. The beach sand removal is prohibited in many parts of the world. Since removal of beach sand will exacerbate erosion leading to higher cost for coastal protection, it should be banned using all authoritative measures.

11.3.5 Reservoir Sand

The reservoirs contain huge volumes of sand, sandy mud, and clay within its basin. Studies reveal that the storage capacity of these reservoirs is reduced considerably due to sediment deposition/siltation. These reservoirs are a good source of construction grade sand. The desilting will also enhance the water storage capacity of the reservoirs. But desiltation has to be carried out only after proper environmental impact assessment by competent scientific agencies.

11.3.6 Marine Sand

In many countries, marine dredged sand forms important component of aggregate supply. In northwestern Europe, extraction of sand from marine sources has taken place for over the past 70–80 years. Dredging for marine aggregate is practiced in countries like the Netherlands, Great Britain, Denmark, Germany, Belgium Japan, Persian Gulf, and France (Kondolf et al. 2002). Marine aggregates are one of the traded commodities in Europe. Considering the environmental concern of land-based sources of sand, the marine sources are to be seriously looked into for construction grade sand. The resource survey, techno-economic feasibility, and environmental impact need to be studied before embarking on offshore mining.

11.4 Sand: Artificial

11.4.1 Crushed Rock Sand (Manufactured Sand)

Crushed rock sand is produced by drilling, blasting, washing, and sieving. The degree to which crushed rock sand can replace natural sand will vary with the rock type, the type of processing, and end use. Also well graded and well-shaped fine aggregate produces concrete with fine finish. Many crushed rock sands have undesirable particle sizes and grain shapes which affects concrete quality and workability. Crushed rock sands are successfully used as fine aggregates in many developed countries. In some cases, the crushed rock sands are blended with natural sands to produce desired grading and workability. Appropriate strategies are to be evolved for promoting rock mining in notified areas, establishing manufactured sand processing units and proper marketing.

11.4.2 Secondary Sand

A variety of waste materials occur in the course of construction and demolition activities. They include materials such as concrete, natural stone masonry, bricks, etc. Many of these materials have the potential to be recycled as secondary aggregates. Concrete and rocks may be crushed and graded to produce secondary

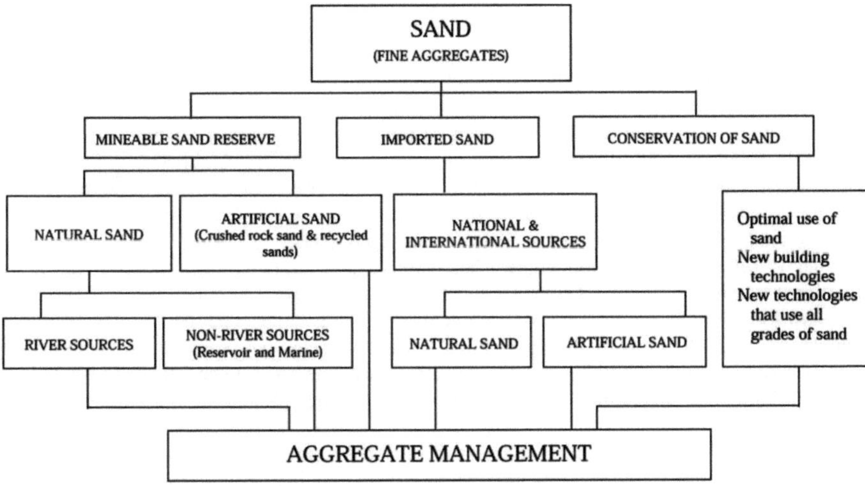

Fig. 11.2 Strategies for aggregate management in Kerala

aggregate, which can be used for construction purposes. In the United States, recycled secondary aggregate contribute about 1 % of the total aggregate demand in the country. Recycling of concrete rubble not only avoids the problems of the disposal of demolition wastes but also reduces environmental effects of new aggregate generation.

11.5 Conservation of Sand

Conservation is another most important method in the wise use and management of sand resource (Fig. 11.2). The following actions are required for the effective conservation and optimal utilization of the limited river sand resources.

- Use river sand only for construction and not for land filling and reclamation.
- Evolve new building technologies with reduced sand requirements.
- Evolve new technologies for the use of all grades of sand in constructions.
- Use alternatives to concrete and cement–sand mix in building technology.
- Introduce penalties for river sand overuse.

A new approach that ensures usage and extraction of river sand at lower regulated levels, use of cost effective alternatives to river sand, reforms in the existing building technologies and conservation of river sands, and above all, a new attitude toward building/house constructions is required to reduce the gap between demand and supply in construction industry on one hand and environmental protection on the other.

References

Allender R, Hollyer SE (1972) The sand and gravel resources of the area south and west of Woodbridge, Suffolk. Assessment of British sand and gravel resources, No. 3, Report No. 72/9, Institute of Geological Sciences, Natural Environment Research Council, London, p 12

Jensen ML, Bateman AM (1979) Economic mineral deposits. Wiley, Chichester, p 593

Kondolf GM, Smeltzer M, Kimball L (2002) Freshwater gravel mining and dredging issues. Report prepared for Washington department of fish and wildlife, Washington department of ecology and Washington department of transportation, p 122

Padmalal D (2011) Construction grade sand: sources and strategies. In: Proceedings of the national seminar on mining of river sand and its impacts on the environment, 18–19 Feb 2011, Centre for Water Resources and Management, Kozhikod, India, p 55–64

Padmalal D, Arun PR, Maya K, Ramachandran KK (2004) Mining of some natural resources of greater Kochi region—problems and perspectives. In: Rabindrakumar GR, Subhash N (eds) Silver jubilee compendium. Centre for Earth Science Studies, Thiruvananthapuram, pp 347–357

Scott PN, Eyre JM, Harrison DJ, Steadman EJ (2003) Aggregate production and supply in developing countries with particular reference to Jamica. British Geological Survey, Commissioned Report CR/03/029/N, Nottingham, p 44

Subject Index

D. Padmalal and K. Maya, *Sand Mining*, Environmental Science and Engineering,
DOI: 10.1007/978-94-017-9144-1,

MIX
Papier aus verantwortungsvollen Quellen
Paper from responsible sources
FSC® C105338

If you have any concerns about our products,
you can contact us on
ProductSafety@springernature.com

In case Publisher is established outside the EU,
the EU authorized representative is:
Springer Nature Customer Service Center GmbH
Europaplatz 3, 69115 Heidelberg, Germany

Printed by Libri Plureos GmbH
in Hamburg, Germany